What People Are Saying About Louise Jupp and *Precision Farming From Above*

This book is up-to-date, informative and to the point. It mirrors what the author is attempting to achieve for farmers who realize the importance of employing the precision farming methods described.
Louise has an answer.

Duncan Scott Director,
Terreco Environmental, South Africa

Louise Jupp has provided a highly useful and practical guide for farmers looking to incorporate drones into their farm operations.

The book explains how drone technology offers the versatility to improve land management and increase crop yield plant by plant rather than the traditional field by field basis.

Precision Farming from Above is a must-read handbook for any farmer or grower seeking to operate a more profitable agribusiness.'

Sharon Rossmark CEO *Women and Drones USA*

I personally met Louise and I appreciated her commitment, not only for her work, but also for a better world.

In fact, agriculture is essential to provide food to billions of humans (and also other products.) But, to do so, agriculture requires resources which are limited (soil, water) or which constitute a threat to environment (energy, chemicals).

In this book Louise explains, in terms understandable without the need of being experts, how drones may help to improve quality, productivity and economic efficiency, while minimizing the use of resources.

Professor Filippo Tomasello, Senior Partner
EuroUSC-Italia, Italy

Louise Jupp has perfectly captured how drones can reshape the future of Agribusiness. Optimizing yields whilst managing costs. It's brilliant!

Purdy Patel Private Banking, *Standard Bank*, South Africa

Rarely, if ever, in the past twenty-odd years, have I come across a person with such raw passion and inquisitive approach about a subject that I make a living from. Drones offer us so much, and the specific and clear focus on the agricultural applications of these innovative tools makes for incredibly revealing and stimulating reading.

This book is a must-read if you are interested in drone applications in the agricultural space: basically, the future.

It unravels some of the technological mysteries surrounding these machines and makes a compelling case to get involved in this exciting market.

Read this book and you will be informed and empowered to succeed in the high-tech precision farming world, while learning about fundamental aviation safety and regulatory principles – an exceptionally difficult balance made easy through this well researched and brilliantly articulated text.

Well done, and thank you Louise!

Rob Somers *22 Degrees South Aviation*, South Africa

Louise Jupp's book provides a ground-breaking opportunity for progressive farmers keen to harness innovative techniques: aiming not at nearby horizons but way above the treetops, where only airborne surveillance can provide the eagle-eye precision necessary for expert planning.

Drones provide a powerful new force for farming, so alight before it's too late. I can't wait to see this versatile approach expanded into geotechnical engineering operations.

Dr GV Price Geotechnical Specialist, South Africa

In the two decades I have known Louise, I have never failed to be surprised, impressed, humbled, and entertained.

She has an expansive knowedge-base, with a passion for natural history and has the capacity to become deeply immersed in any field she chooses - as she has in the past four years within commercial drone technology and its potential influence and application in agriculture.

I cannot begin to express my delight at the publication of this book, which represents many years of sacrifice, sunburn, intensive study and endeavor on her part.'

Dr Mandy Uys *Laughing Waters & Associates*, South Africa

Terreco are truly ahead of their time in the field of *Environmental Services*. In modern business and projects there is a serious under-utilization of technology to increase productivity and benefit.

Terreco have proven time and again that they are prepared to push the boundaries of what can be achieved, and through hard work and the use of modern tech, they have become market leaders in the South African and African markets.

Bruce Fennessy Managing Director
Auric Consulting, South Africa

What an interesting, dedicated and a knowledgeable individual in her field Louise is. Since I started contracting to Terreco a few months back we have built an amazing professional relationship. I wish her all the best.

Barry Dean Delport, B*arry Dean Consulting,* South Africa

PRECISION FARMING FROM ABOVE

How Commercial Drone Systems are Helping Farmers Improve Crop Management, Increase Crop Yields and Create More Profitable Farms.

Louise Jupp

WRITING MATTERS PUBLISHING

Precision Farming From Above: How Commercial Drone Systems are Helping Farmers Improve Land Management, Increase Crop Yields and Create More Profitable Farms.

First published in August 2018

Writing Matters Publishing (UK)
info@writingmatterspublishing.com
www.writingmatterspublishing.com

ISBN 978-1-912774-06-7 (Kindle)
ISBN 978-1-912774-07-4 (Also available in Pbk)

Copyright Louise Jupp

The rights of Louise Jupp to be identified as the author of this work; have been asserted in accordance with Sections 77 and 78 of the Copyright Designs and Patents Act, 1988.

A CIP catalogue record for this book is available from the British Library.

All rights reserved. No part of this book may be reproduced in material (including photocopying or storing in any medium by electronic means and whether or not transiently or incidentally to some other use of this publication) without the written permission of the copyright holder except in accordance with the provisions of the Copyright, design and Patents Act 1988. Applications for the Copyright holders written permission to reproduce any part of this publication should be addressed to the publisher.

Disclaimer: *Precision Farming from Above* is for information only.

It does not replace the advice of professional commercial drone surveying operators and other professionals in the agriculture sector. The reader is advised to seek professional advice before implementing any of the insights they might gain from reading this book or before undertaking any of the described activities using drones.

The views and opinions expressed in this book are those of the contributing authors and do not reflect those of the Publisher and Resellers accept no responsibility for loss, damage or injury to persons or their belongings as a direct or indirect result of reading this book.

All people mentioned in case studies have been referenced and/or used with permission; and have had names, genders, industries and personal details altered to protect client confidentiality and privacy. Any resemblance to persons living or dead is purely coincidental.

All images have been used with permission.

Back cover Photograph: Madeleine Chaput
Cartoons: Andrew Priestley

Contents

- 9   Note To Readers
- 11   At a Glance
- 15   Overview: Precision Farming From Above
- 17   Introduction
- 31   Part 1: Farming Today
- 43   Part 2: What I See That Works For Farmers.
- 49   Part 3: Having Better Quality Information
- 71   Part 4: Spend More Time On Productive Tasks
- 83   Part 5: Creating Better Management Strategies
- 93   Part 6: The 5 Step Precision Farming Blueprint to Smarter Farming
- 105   Part 7: Yes, But Can I Do This Myself?
- 125   Part 8: Making It Happen
- 133   Case Studies
- 137   References
- 143   Bibliography
- 147   Glossary
- 149   Acknowledgments
- 151   About the Author
- 153   Contact

Note to Readers

For Readers in South Africa

Please note, at the time of printing, we are finalising our drone operating licensing and permits with the South African Civil Aviation Authority (SACAA).

Louise is already a licensed drone *pilot* with over 300 flying hours. She was one of the first women to obtain her pilot license from the *South African Civil Aviation Authority* in 2016.

Terreco Aviation's intention is to be the premier provider of *commercial drone* surveys in South Africa. We will commence providing a full range of drone services once we have completed the licensing process in 2019.

For reasons outlined in the book, you should only deal with licensed pilots and service providers.

For Readers Outside South Africa

We can now, however, quote and provide our services in jurisdictions *outside* South Africa.

Precision Farming From Above Is For Information Only

This book *does not* replace the advice of professionals within the agriculture sector.

The reader is advised to seek professional advice before implementing any of the insights they might gain from reading this book or before undertaking any of the described activities using drones.

For reasons outlined in this book, you are cautioned against using unlicensed, amateur service providers using *recreational drones*.

Be advised that you may be held as liable as the operator in the event that you have breached civil aviation laws, public safety or both.

Wherever you are based, note that infringing airspace laws carry stiff penalties for justifiably good reasons.

At a Glance

Extensive Review

Precision Farming from Above is the result of an extensive review of books, reports, webinars and websites produced by international leaders, software developers and manufacturers within the drone industry as it relates to *commercial drone* aerial surveys, *precision farming* and agribusiness development, as distinct from *recreational drones* used by the general public.

The term *drone* is inclusive of *unmanned aerial vehicles* (UAVs), *unmanned aerial systems* (UAS), *remotely piloted aircraft* (RPA), *remotely piloted aircraft systems* (RPAS) and *remotely piloted vehicles* (RPV).

The book explores the use of sophisticated pilot-operated, *commercial drone surveying systems* as applied to agriculture.

It includes:

- Drone technology - recreational vs commercial
- Aerial surveys for agribusinesses
- High resolution cameras
- GPS guided flight programming software
- Data analysis
- Application

Importantly, I will include:

- Pilot licensing
- Civil aviation regulations
- Compliance

I especially want farmers – globally - to understand the enormous benefits *commercial drone surveying systems* offer to their businesses as it relates to *precision farming*, productivity gains and more sustainable land and crop management.

Precision Farming from Above explains how this is achieved through the fundamental combination of:

- The versatility and manoeuvrability of the aircraft that carry the cameras;
- The variety and sophistication of cameras carried by the drones;
- The programmable flight planning software for accurate surveying and photography;
- The analytical power of the online software platforms;
- The comparatively rapid conversion of processed information into actions for farmers to apply.

The book also explains the massive advantages in turnaround time for this range of high-quality information to be converted into actionable data for farmers. This can be anything from minutes to 48 hours, and into a variety of compatible formats.

This high-tech, precision farming approach means farmers have a near real-time understanding about the state of their farms and are therefore more able to make more informed, proactive and effective decisions on their current and future activities.

They are also able accurately to operate on a *plant-by-plant* basis because of the resolution of the data, rather than the traditional *field-by-field* basis. This translates into achieving more efficient and sustainable operations, with the many benefits to time, budgets and management strategies this offers.

Who Should Read this Book?

Precision Farming from Above is specifically aimed at progressive farmers who have medium to large-sized operations that they may own or operate as individuals or as part of cooperatives.

In any case, *Precision Farming from Above* is relevant to any farmer anywhere in the world who is looking to improve land management, increase crop yields and operate a more profitable and environmentally sustainable agribusiness.

In this book, the term *farmer* is inclusive of anyone growing crops for commercial reasons.

Three Predictable Problems

While farmers face many problems, their productivity and profitability is consistently disadvantaged by three recurring and predictable factors:

- The quality of information they have about the state of their crops, soils or structures.
- Being tied to mundane time-consuming tasks.
- Inefficient management strategies.

Commercial drone surveying, high grade data and the smarter, precision farming they introduce are providing solutions to these issues.

References and Glossary or Terms

Readers will find an extensive list of references and a glossary of terms used at the back of this book.

About the Author

Louise Jupp has a Master's Degree in Environmental Science.

She has over 26 years' experience in environmental management in the UK, Europe and Western, Central and Southern Africa.

She co-founded *Terreco Aviation (Pty) Ltd* with her business partner in 2016.

Her goal is to help farmers and growers world-wide achieve more profitability and financial security in their agri-business operations, and to do so in an innately more sustainable way.

Louise is a licensed drone pilot with over 300 flying hours. She was one of the first women to obtain her license from the *South African Civil Aviation Authority* in 2016.

Overview:
Precision Farming From Above

Introduction

I have friends who are farmers in South Africa. Most have been farming for several generations. They are passionate about their families, their farms and their livelihoods.

However, I worry about the future of farming, and their future.

Farming is increasingly more demanding and stressful. Many factors that are associated with growing good quality crops and generating secure incomes, are unpredictable.

Too many things can go wrong, or are beyond their direct control, such as droughts, freak weather conditions, or a weakening currency.

A good harvest can be as much a product of luck as good planning and management. And yet, a strong national agricultural base is essential for all developed and developing countries.

Farming Predictions

Many nations are currently producing enough food to feed their populations. But studies predict global agricultural production must substantially increase to respond to the expected global population growth.

A *PwC* report (2016) states global *'aggregate agricultural consumption will increase by 69% from 2010 to 2050. This increase will be mostly stimulated by population growth from 7 billion to 9 billion by 2050.'* [1]

The predicted populations of the Middle East and Africa are expected to be 3.4 billion alone. This is *'likely to be more than the populations of China and India combined.'* [2]

Specific to South Africa, a *World Wide Fund for Nature* (WWF) report predicts food production or imports will need to *'more than double'* if it is to feed a population that is expected to increase from 49 million in 2009 to 82 million by 2035. [3]

Farming Is a Business

Farmers know they play a critical role in feeding a nation, driving trade, contributing to the gross national product, providing employment and supporting the industrial growth of nations.

But the reality is: *farming is a business.*

Farmers world-wide must still generate a steady income to provide for their families, pay employees and maintain a profitable successful business.

And most want to pass a successful family business on to the next generation and continue a proud lineage that typically already goes back several generations.

All of this takes place within the global reality of a wide range of factors, including: changing weather patterns; changing global markets and financial drivers;

more environmentally aware consumers; and a growing need to utilise natural resources more conservatively, to name but a few.

Under these circumstances, farmers are looking for anything that will give them the *edge* to maximise their yields, maximise their financial returns and maintain a stable business from one year to the next.

Such an *edge* would help farmers reduce their vulnerabilities to the vagaries and uncertainties that are inherent to food production processes.

Superior Knowledge Gives You The Edge

Some farmers are already responding to these factors and achieving the *edge* they want.

The most successful ones are adopting more effective farming methods that enable them to compete more successfully with greater economic certainty and purchasing power.

Specifically, they have achieved this *edge* because they have *superior information about their farms*, and, as a consequence, they have better management strategies in place to enable them to be more productive.

Importantly, they are better placed to increase their yields and to do so in a sustainable and predictable way.

However, this is *not* the norm.

I See The Opposite

I see farmers who struggle from one season to the next, often due to no fault of their own, but because they are vulnerable to factors currently viewed as outside their immediate control.

They are vulnerable because they don't have enough high-grade information about their farms.

They spend valuable time on mundane activities. And they have no truly dependable strategy for the oncoming season.

The reality for many farmers is that they carry a growing debt burden and the real risk of bankruptcy if their harvest does not meet expectations.

But there is hope.

Precision Farming

The same major analytical reports describing the need for global agriculture sectors to rise to the challenge of feeding an expanding population also conclude that farming practices are radically changing.

Commercial farmers are already embracing *smart* or *precision* agriculture in order to become more efficient, more productive and more sustainable.

The key to smarter agriculture is having more and better quality information about the farm.

There are already a variety of *smart* farming techniques and technological innovations available to farmers including *smart* tractors, combine harvesters and chemical delivery systems.

I see that *commercial drone surveys* are increasingly a key part of the solution and a method for bringing the benefits of *precision agriculture* to a greater number of farmers.

Aerial surveys using sophisticated pilot-operated, commercial drones, high-end software and detailed analytics are fast becoming an essential tool in the farmer's toolbox.

They are as fundamental as a tractor.

Commercial Drone Surveys

Drone surveys give impeccable high-resolution detail that surpasses what is available via traditional fixed wing or helicopter surveys.

In most cases this information is immediate, safer to obtain and significantly more affordable and versatile.

Around the world, access to *smart farming* techniques through professional drone surveying is giving farmers real, powerful and immediate opportunities to gain control over the unpredictable, to improve their circumstances and to achieve their goals.

To paraphrase the analysts, drone surveys are increasingly part of modern farming because they are transforming a traditionally data poor industry into one where farmers can, for the first time, make informed crop management decisions based on real-time data. [4]

Agriculture world-wide is *already* benefiting from professional drone surveying. For example, *'drought assessment, plant stress monitoring, crop health monitoring, yield monitoring and field surveying before planting.'*

More farmers are starting to realise the potential value of professional drone surveys in providing rapid, high quality intelligence.

Recreational vs Commercial Drones

Drones have military origins, but civilian uses are now far outstripping their military applications.

Be clear, I am *not* talking about small light weight, *recreational drones* that show up in the toy or electronics section of a department store. The demand for and ease with which *toy drones* can be purchased and used by all ages compared to just five years ago is staggering.

Drone suppliers would have you believe that anyone can purchase a drone and expect to fly it as soon as they've finished reading the *Quick Start How-To Guide!*

However, the major drone manufacturers, such as *DJI*, are moving away from the recreational market and have created industry-specific *commercial drones* and accessory bundles in recognition of the enormous commercial value and applications of their products.

I am talking about sophisticated, commercial, industry-standard drone systems that can only be operated by properly trained, qualified and licensed pilots who must comply with civil aviation requirements.

Furthermore, the information collected via specialist cameras and processed using proprietary software must be analysed and interpreted to maximise the full value of the data for agricultural applications. In other words, the full value of a commercial drone system is not attributed to just the drone or the camera or software but to the integrated system comprising all these features.

There is more value beyond the usual perception that drones are just for taking nice aerial photographs of farms.

Commercial Drones

Commercial drones, in their dominant multi-rotor configurations, typically quadcopter, are largely the result of a convergence of technological advances, including those associated with electric motors, GPS, batteries and smartphones. This led to the initial development of drones which were lighter, easier to control and manoeuvre in flight, and able to carry high-grade cameras that provide real-time data flow.

Operating software for the more sophisticated models (e.g., the *DJI Phantom 4 Pro* series) includes flight management, obstacle avoidance, autonomous flying and fail-safe features that reduce the inherent risk of crashing.

Tablets or smartphones can be used to fly the drones, providing a heads-up display for the pilot with a traditional radio controller.

Data, in the form of photographs and videos, are stored on board the drone using microSD cards. Alternatively, you can *live-stream* footage directly to other devices.

Toys or Tools?

Perhaps you have seen small drones hovering above neighbouring properties, buzzing about at local events, gliding through picturesque areas, or even zipping through obstacles at breakneck speed in televised drone races.

Alternatively, you might have seen quirky news stories about pizzas being delivered by drones, or *Amazon* considering the delivery of their parcels by fleets of drones.

You should know that in most countries, as soon as a drone is used for a *commercial application, it is no longer deemed a toy*, and it comes under stringent aviation laws.

This is especially so in South Africa.

PRECISION FARMING FROM ABOVE

While some farmers have realised the potential drones offer to their business, there is a pervading perception that drones are a gimmick, a *toy* used for playful hijinks, taking nice photos or videos and even for spying on neighbours! (Which, by the way, is highly illegal.)

I frequently hear such comments from farmers when we display our drone services.

This is an unfortunate perception that prevents farmers from benefiting from the true potential of commercially applied drones. They are *tools*. More than that they are management tools.

Commercial drone surveying is a professional service utilising sophisticated equipment and software and reliant on a trained and experienced, *and licensed,* team.

If you're only seeing the value of drones from the recreational perspective of drone racing or just taking pretty pictures, then you're seriously missing a significantly bigger picture.

Worse, you're missing out on the immense benefits commercial drone surveying offers to your farming business.

This is one reason why I have written this book.

I want to break the myth that drones are just *toys*.

I want to make clear distinctions between *recreational drones (toys)* - and *commercial drones (tools)* and in doing so, explain how *commercial drone systems* can be applied to great advantage for modern agriculture.

I am referring to the use of high tech, highly sophisticated *commercial drones* – technically *aircraft* - that are subject to the stringent rules and regulations set and managed by international and national civil aviation authorities.

The *same* rules and regulations that typically:

- Consider drones a category of *aircraft;*
- Define commercial and private (or recreational) uses of drones;
- Require drone operating companies to comply with the *same* requirements as manned aviation operators;
- Describe the manner in which commercial drone operations may be safely carried out in defined airspace by *licensed* commercial operators;
- Require operators to be regularly audited for performance and compliance;
- Require full insurance cover to operate;
- Require drone pilots to obtain a *Remote Pilot's License* from approved training schools after completing a theory examination and flying test.

All of the above are certainly true of the South African and European requirements.

Licensed Drone Pilots

In South Africa, commercial drone companies or operators and their pilots are legally held accountable for their activities to the *same* extent as manned aviation operators and pilots.

For example, a licensed drone pilot, such as myself, will suffer the same consequences as a commercial manned aviation pilot if found to have had an alcoholic drink eight hours before flying!

We are also legally required to operate safely and provide clients with a professional service.

In short, there is a world of difference between responsibility and accountability for commercial drone *pilots* (and operators) compared to recreational pilots and hobbyist drone users.

Thankfully, this perception about drones as toys is rapidly changing.

The value of *commercial drone* services across many commercial sectors is being explored, recognised, developed and exploited – and, I emphasise, including *agriculture*.

There is no question that, globally, commercial drones are becoming indispensable business management tools and are rapidly transforming those businesses that have adopted the use of commercial drone systems. [5]

Agriculture is the second largest drone market that will benefit from commercial drone-based services, after infrastructure.

The global value of commercial drone services to agriculture alone is already estimated to be in the order of USD$32.4 billion. This represents 25% of the total global value of drones across multiple markets. [6]

If You're a Farmer, Then This Book Is For You

If you are ready to become a more successful farmer, then this is the book for you. I've written this book to show you what commercial drones and drone services can do for your agribusiness.

Specifically, I describe:

- Why commercial drone systems, incorporating supporting software and specialist high quality cameras, are invaluable to *your* business.
- How commercial drone surveying is one of the most important tools you can and must integrate into *your* farming operations, especially if you want to acquire that edge towards being more financially successful.
- The types of data *you can collect*.
- How easy it is to collect a range of invaluable information using commercial drone systems; and
- How *you* can use this information to increase your yields, profitability and environmental performance.

I also add references to case studies from around the world to demonstrate how and why commercial drone services are essential management tools that you must consider utilising in your business.

To get the most from this book, I would encourage you to critically review situations you encountered during your last growing season.

You might have wished you had more warning about a specific problem that had occurred with your crops or soils.

Or a mundane, time-consuming task that kept you away from an essential activity which had knock-on effects.

Or you felt as if you were just reacting to one long costly crisis management situation.

Consider how these problem areas affected your *top*

and *bottom-line* and whether the solutions you've applied to date worked as well as you really needed at the time, assuming you were able to monitor their performance.

Compare your experiences against the information presented in this book and ask yourself what you might have achieved had you used professional drone services.

How To Get The Most From This Book

This book has been specifically written for farmers with medium to large sized farms growing crops, including grains, root crops, sugar cane, coffee and tea, hops, pastures, plants for seeds and oils, nut and fruit orchards and vineyards.

I have focused on South African farmers, but everything I present in this book has global relevance for farming operations of all sizes.

The term *farmers* equally refers to *growers* throughout the book.

Additionally, any reference to *crops* or *plants* is equally interchangeable with any produce grown for consumption in one form or another.

Similarly, where I have used the terms *fields* or *farms*, these can be exchanged for orchards, vineyards and plantations.

Finally, the term *commercial drone systems* collectively refers to the essential combination of:

- The aircraft (or drone);
- The specialist camera(s);
- Flight planning software;
- Data processing and analytical software used to generate actionable data.

The true value of the commercial drone service to agricultural businesses is the sum of all these components especially when operated by highly-trained and licensed service providers.

If you are prepared to integrate the high-tech services described in this book into your business, then you can achieve better yields and more financial success.

This will result from having:

- Better information on what's happening on your farm;
- Using your time more productively; and
- Being able to build and implement better management strategies for maximising your resource efficiency and your farm's productivity.

Whatever your current views on drones, commercial drones are already here.

Professional drone services are already delivering immense benefits to farmers world-wide. I know that the information delivered by commercial drone surveying to farmers is a game changer.

My goal is to clearly explain these benefits to you so you are left in no doubt about the value of commercial drone surveying *for your business.*

By the time you've completed this book you will know more about what drones can do for you to the extent you will be asking yourself, "*Can I afford not to use them?*"

Part 1
Farming Today

Part 1: Farming Today

World-wide, farmers carry huge responsibilities, whether they are producing food for their families or growing food products for domestic or international mass consumption in one form or another.

Today's farmers still face constant challenges growing their produce (whether grapes, grains, fruits and nuts) and maintaining their businesses despite all the advances that have been made in modern farming. Some challenges are inherent to the natural environment on which all food production is dependent, on market and political forces, or are created unwittingly by the farmers themselves.

Many challenges for any farmer are external to their own actions and include forces that influence the growth of good, high quality yields as well as the factors that influence the value of the yield once the cost of resources used etc are taken into account.

But ultimately, the growing environment, influenced by the climate, weather, soil conditions and prevalence of bugs and disease, has the greatest impact on their fortunes and is the root cause of most problems.

Three Core Problems

I have identified three core problem areas that I believe will repeatedly hamper your ability to respond effectively to the natural challenges you face and manage your productivity, including:

- A lack of accurate information about your crops or soils at the right time;
- Having insufficient time for productive tasks;
- Inefficient or compromised management strategies as a consequence of paucity of data and time.

Unfortunately, these three core problems are entrenched within many traditional farming practices. Therefore, unless you are able to resolve these problem areas, your business will remain vulnerable to your being: typically, ill-informed about the actual state of your crops and soils; tied up with time-consuming, inevitably mundane tasks, and; compelled to manage each problem as a crisis.

Problem 1: Lack Of Accurate Crop Information At The Right Time

Using traditional practices, *'it can be difficult or impossible for the farmer to react to a problem like a disease outbreak before it's too late, or the costs to treat it have soared.'* [7]

How many times have you been in an *if only* situation about the state of your crops and soils because of the lack of information or late availability of information?

One of those *if only* you'd known sooner you could have made a plan to recover the situation whether it would have been timely respraying, re-planting or repairing equipment, such as a pivot.

Traditionally, the state of crops and soils is determined by walking and inspecting the fields, vineyards and orchards and taking selective samples for testing. Walking the fields of any sized operation takes time and, in most cases, the area covered during one of these walks will represent only a fraction of the total area that has been planted.

Additionally, views from ground level, next to or within the crops, vineyards or orchards are not going to provide an area-wide understanding of the state of the plants. Finding some problems will be more by chance than by design.

Alternatively, you may choose to drive your property, fields, orchards or vineyards partly to cover large areas or because your time is limited (see the next section). But, driving along at any speed at a distance from your plants is still only going to give a snapshot of the condition of your plants at best. Again, you are not going to see into the centre of the field, orchard or vineyard, especially once the plants start to grow.

As the PwC report makes clear, *'Vast fields and low efficiency in crop monitoring together create farming's largest obstacle. [Crop] Monitoring challenges are exacerbated by increasingly unpredictable weather conditions which drive risk and field maintenance costs.'* [8]

More effective area-wide inspections can be made using remote sensing data acquired by satellites or commissioning aerial surveys using aircraft, where budgets permit. But, these methods are either slow in returning the information or costly, and usually both.

Satellite data covering the farm can be purchased but there are limitations to the availability, flexibility and resolution of this data. *'Until recently, the most advanced form of monitoring used satellite imagery. The main limitation was that images had to be ordered in advance, could be taken only once a day and were not very precise.*

In addition, the services were extremely expensive and gave no guarantee of quality, which could easily drop on a cloudy day.' [9]

Additionally, satellite data only suit very large farm operations because of the resolution and field of view.

Aerial surveying by manned-aircraft may be more flexible than satellites, but it is an expensive exercise and, assuming you can afford this option, is typically best used for large farms. The turnaround time for processing the information can be time-consuming and complicated by the vast amount of data collected for larger farms. This means the information can be out of date by the time you can use it. However, the use of aircraft may not be an option for your location, or it is simply not a cost-effective option for your operation.

Finally, with the exception of aerial surveying or satellite data, information that you will physically collect on foot or via your vehicle about the condition of crops and soils is limited to what you can physically see. Some information about plant health is simply not visible without the use of specialist equipment or specialist services.

Other business sectors would be significantly less tolerant of their managers making strategic company decisions on this level of information!

And yet, up until recently, the quality and value of the data collected on the physical state of your crops and soils meant you were routinely vulnerable to: ending up with lower yields; nasty surprises at harvesting; costly remedial measures; lower returns; and more financial stress, to name but a few consequences.

Problem 2: Insufficient Time For Productive Tasks

Your primary goal is to produce good quality crops, so you can generate a good return on your investment.

Given the amount of money literally ploughed into a season, you would want to spend as much time as possible making sure that your plants get the best attention, water, chemicals and fertiliser they can.

But how much time do you really have?

If you are like the farmers I know, you never seem to have enough time in a day.

You are always rushing backwards and forwards across fields or between orchards and vineyards for one reason or another.

If you want to inspect crops, soils or infrastructure, the sheer size of your farms and the distances between fields, vineyards, orchards or plantations results in your spending a lot of time travelling.

In any one day, you spend more time travelling than being productive and getting on with the important business of growing crops.

Some farmers may only have an opportunity to complete a visual scan of their property once a month, and then only by driving their boundaries. Driving on bone-shaking dirt roads or *jeep tracks* across the property makes for slower trips and further time delays, especially when the weather is poor, and the roads become mud wallows.

Travelling to and from the nearest town for supplies and resources can also take a fair slice out of your day, depending on the distances involved and the quality of the roads. Not only do you have limited time during the day, but your time rarely seems to be yours to control.

There is so much you need to achieve in any given day in preparation for and during a growing season, and there is always a spanner thrown in the works from somewhere.

A pivot is blown over during a sudden squall, a fire threatens the general area, staff do not turn up for work as expected, there is a mechanical fault on essential machinery, or you simply have a puncture.

Simple activities have a habit of rapidly turning into all-day affairs for a variety of reasons - planned or unplanned.

One way or another, you must divert your attention away from your original task to deal with a problem. And then there are those situations where an activity (or problem) just takes more time than you thought it would through no fault of your own.

The tragedy is you may end up being distracted by these time-consuming but ultimately mundane issues and, worse, having your attention diverted away from the essential issues. This situation is too common.

It is inevitable that, without enough available time to dedicate to essential tasks, you are at risk of missing opportunities to identify and address problems on your farms, to give your problems 100% of your attention, or to complete tasks within the time frames that will best maximise yields and productivity.

Problem 3: Inefficient Farm Management Systems

Farming gives me the impression of being a bit like the lottery. You never know how many of your numbers are going to come up and if you're going to win the jackpot, collect a small sum or get nothing.

This being said, you are investors.

Each year, you invest time, money and resources (hearts and souls!) into businesses with the expectation of recovering costs and, better still, making a profit from the investment. For those who operate from a position of debt, you are aiming to reduce the amount owed with a good season.

To complicate matters further, you invest within a risky environment; one that includes fluctuations in demand and currencies, changing markets and weather patterns. Just these factors alone will change the price of essential items such as the seeds and cultivars; water, fertilisers and chemicals; fuel and electricity; machinery and parts; as well as the value of your harvests. The effect of these rising costs on your business are compounded by lower market prices for yields as well as any loss of yields.

Rising costs and lower returns will leave anyone lying awake at night wondering what they can do to turn their fortunes around.

Under these circumstances, you would want to make the best plan possible for a new season to better manage your resources, labour and spending in order to produce good crops, improve yields as well as maximise your return on investments – just like any other business sector.

And yet it is difficult for you to forward-plan with any degree of certainty, or even be able stick to your budget or work plans, because of the unpredictable nature of your operations within the natural environment.

One episode of scorching weather can wreck plans

and force replanting. There are so many unpredictable scenarios that could unfold in any given season that it is almost impossible to plan courses of action or maintain enough funds for contingencies.

It goes without saying, the smaller the farm operation, the greater its vulnerability to rising operating costs, unplanned expenditure and lower returns.

Furthermore, smaller farms may not be able to build a buffer or contingency for weathering the leaner seasons.

Once a farming business becomes vulnerable there are knock-on effects which, for example, may drive you to selling products when prices are low out of financial necessity. You are unable to hold onto your produce until prices improve because you do not have the financial freedom to do so.

It also follows that banks tend to behave more conservatively and rigidly towards smaller businesses.

Given these financial situations, it is obvious you will try to control what you can within your operations by reducing costs wherever possible or being cautious about spending money at the same time as trying to maximise the value of your forthcoming harvests.

While great strides have been made to help reduce the inefficiencies and waste in modern farming, many farming methods are still wasteful. For example, you will generally respond to a nutrient issue by spraying an entire field. This means fertiliser will be applied to areas that do not need additional nutrients. Not only will you waste the fertiliser, but you will use more fuel in the machines used to apply the fertiliser, as well as add a wear and tear cost to your vehicles/machinery. You or your team will also waste time spraying the entire field when there is no need.

When it comes to maximizing yields, any growing season may be affected by uncharacteristic or unexpected weather conditions such as droughts, too much rain, early frost, mild winters or hail storms.

Then there is the risk of diseases attacking crops, or pests eating or damaging the plants, flowers or fruits.

Any one of these events will require actions and resources that add to your woes, if budgets are tight or harvests are less successful.

Some losses experienced during a season (such as fire or hail damage) can be covered by insurance, but claims take time to resolve and for the monies to be paid. This situation may mean you do not receive the monies in time to act at the right time. However, if you cannot afford to take out insurance in the first place, then this safety net is absent.

So What Does This All Mean To You?

These three core problems actually have a cumulative effect which can leave you exposed to the risk of bankruptcy should you experience a particularly poor harvest and/or a lower income for any single season or over a consecutive number of seasons.

I admit I have painted a bleak picture, but I have seen these scenarios unfold. They unfold because you simply do not have enough good quality information, time or management strategies that will help: maximise your yields; improve your financial situations; create room for financial buffers or contingencies; and give yourselves a better footing from which to start the next season.

Inevitably, like the farmers I know, you are stuck with limited options mostly because of the financial restrictions you tend to carry from one season to the next.

Still, you continue to do the best with what you have and,

somehow, you always manage to make a plan to produce your crops, pastures, grapes, fruits and nuts and survive as a business. Just.

But it's always a battle and you typically remain cornered by financial constraints even with a successful harvest.

This is a difficult cycle to break, especially where you have limited options to absorb additional financial demands. As a consequence, it is hardly surprising that we hear a resistance to change when we speak about this new approach to farming. No matter how this resistance may be justified or expressed, the results are the same.

A *better-the-devil-you-know* scenario prevails where many will rather carry on business-as-usual even though it may perpetuate inefficient practices, waste resources and money and leave you open to ongoing risks.

It's a vicious circle.

The Burning Question

If you could have one thing for your farming business, which would solve all your problems, what would it be?

Alternatively, if you were speaking to the most successful farmer in your line of agribusiness and wanted some solid advice, how would you finish the question, *'How do I ...?'*?

I know you want to produce good quality yields with fewer complications and more efficient use of resources. You want to make more money than you invested (and owe) and become more financially resilient.

I've seen farming publications referring to farmers wanting to make painless profits.

Based on the stress I have seen farmers and growers suffer, I think this is a perfect all-encompassing expression – you just want to have pain-free, sustainable businesses.

I have *an* answer.

Part 2
What I See That Works
For Farmers

Part 2:
What I See That Works For Farmers

If you want a more productive farm and more financial security, then you need:

- More information about your crops, soils and fields;
- To be able to use your time more effectively; and
- A data-supported management strategy in place.

If you have these three factors in place, then typically, your farm's performance improves.

The answer lies in *precision agriculture* or *smart* farming and specifically, commercial drone-supported precision agriculture or smart farming.

The European Parliament defines *precision agriculture* as *'an integrated information- and production-based farming system that is designed to increase long term, site-specific and whole farm production efficiency, productivity and profitability while minimizing unintended impacts on wildlife and the environment.'* [10]

Precision agriculture or smart farming is already having a major positive impact on commercial agriculture around the world in various formats.

This includes GPS guided machinery delivering water, fertiliser or chemicals at precise, variable rates across fields with or without someone in the cab.

However, not every farmer can upgrade their business to this level of high-tech and can therefore miss out on the benefits of this modern form of agriculture.

Drone-supported solutions are the game-changer.

Globally, commercial drone systems have opened up opportunities for farmers at all levels to access and benefit from smarter technology driven agriculture.

Commercial drone systems offer an affordable option for all farmers to incorporate smarter, sustainable farming practices into their businesses from subsistence farmers through to mega farming operators. 'Drone technology will give the agriculture industry a high-technology makeover, with planning and strategy based on real time data gathering and processing.'[11]

Commercial drones combined with sophisticated precision cameras, proprietary software used for precise flying, and the ability to process the imagery and generate meaningful data provides the solution to the three core problems you face.

This high-tech solution ultimately creates the opportunities for you to increase your yields and profitability more efficiently and effectively than currently possible.

The next three parts of this book will explain how each of the core problems we know you face can be solved using commercial drone services, and the cumulative effect each solution provides.

You will see how commercial drone surveying services will:

- change your operations;
- give you the edge to move away from financial uncertainty to a position of financial strength; and
- ultimately give you more business opportunities and freedom of choice.

You will see for yourself how *commercial drone supported* smart farming is an answer to your problems well worth considering.

Part 3
Having Better Quality Information

Part 3:
Having Better Quality Information

'With drones, information is instantly accessible, allowing growers to make immediate decisions.' [12]

Smart farming relies on high quality, real-time or near real-time information enabling farmers to make meaningful, relevant, informed and timely decisions. What sort of information would you really want to have about your crops, soils and fields that you know would give you that edge in producing good quality yields?

- What is the condition of my soils?
- How can I maximise my surface drainage?
- How many seedlings have been planted by my suppliers/contractors?
- Is there a good coverage of emerging crops?
- Are my crops under attack from disease or pests?
- Are my crops in need of more water?
- Are my pivots or irrigation systems working properly?
- Am I over-irrigating or water-logging my crops?
- How many crops did I lose after that hail storm?
- Should I reap my hail-damaged crop now, or leave it to continue maturing?

- Do my crops need a boost of nutrients?
- What are my expected yields, so I can plan my harvesting resources?
- Have the responses I've put in place for a problem worked properly?
- What has been my season-on-season performance per field?
- Are there holes and gaps in my screens and shade cloth covers?
- Are my fences still intact or need repair?

Traditional methods for collecting information to address these questions have been satellite and/or manned aircraft, remote sensing, or simply from *boots on the ground*. However, there are flaws within these options.

Most notably, there is greater opportunity to use commercial drone surveying for smaller fields, sections of fields, vineyards or orchards which is not possible with satellite data, due to resolution or with manned aircraft due to availability, feasibility and costs.

Commercial drone surveying provides the information to answer these questions more efficiently and, significantly, with less delay between collection and being able to take action.

The Versatile Eye In The Sky

Pictures speak volumes.

From the perspective of gathering relevant meaningful data, the commercial drone is just the vehicle or platform for collecting the information.

It is the types and numbers of cameras (or sensors) that the commercial drone can carry, the types of images they collect, the proprietary software used to process and subsequently interpret the processed imagery collected and the resolution of the images, that provides the real value from commercial drone operations on your property.

The Range of Cameras Carried by Drones - Standard Cameras

The range of cameras and sensors that can be carried by drones include a *standard visible spectrum* (or red, green and blue light (RGB)) camera, much like the camera in your cell phone. These cameras tend to have higher specs and are able to record high definition photographs, as well as provide live-feed and record video footage.

Most recreational and entry-level commercial drones are fitted with a standard high definition visible-spectrum camera as part of the normal configuration. These cameras are fixed and typically cannot be switched for another camera or sensor. This is significant because a visible-spectrum camera has limitations on the types and versatility of the information available from the imagery collected compared to a multi-spectral camera.

The standard visible-spectrum camera provides a live view of the areas being over-flown irrespective of whether video footage is being recorded or photographs are being taken. This live view is displayed on a tablet or smart phone that is connected to the remote control as part of the *heads-up* display. This view is important for the pilot to help orientate and navigate, but the same images can be simultaneously streamed in a live feed for you to view at the same time as the pilot via your laptop, tablet or cellphone. This gives you an opportunity to take part in the inspection and guide the pilot or camera operator to specific areas of interest before videos or photographs are captured.

The Range of Cameras Carried by Drones – Specialist Cameras

Other cameras that can be used with commercial drones include *multi-spectral* cameras, *thermal imaging* cameras and *hyperspectral* cameras.

These specialist cameras are more sensitive to different parts of the electromagnetic spectrum than the normal *visible* spectrum cameras we are familiar with.

They capture images and information that is *not* visible to the naked eye. Arguably, they provide the most important information that farmers can use to their advantage.

Specialist cameras are used with more exclusive commercial drones which can either carry several cameras at the same time or have interchangeable mounts that allow one camera to be substituted for another.

The specialist camera most commonly used for precision agriculture is the *multispectral camera*. However, *thermal imaging camera*s and *hyperspectral* cameras are beginning to feature more as they are miniaturised, resolution improves, software is developed, and prices drop.

There are also highly sophisticated *LiDAR sensors* (Light Detection And Ranging) which have been used in agriculture in the past. These sensors use lasers to survey and measure landforms for very detailed and accurate high-resolution 3D modelling.

LiDAR sensors have been carried on aircraft and by vehicles in the past, but smaller sensors have now been developed for commercial drones. Currently, they are very expensive. I will stick to talking about the *multispectral* and *thermal imaging* cameras in this book as these are more widely used for agriculture.

The Number of Cameras Carried

The most popular configuration for mainstream commercial drones is to carry one visible-spectrum sensitive camera only.

More exclusive commercial drone systems provide for two cameras to be carried and operated at the same time. Clearly, commercial drones which can carry more than one camera at a time or are interchangeable have the advantage over single camera drones. This advantage relates to the greater quantity and range of information that can be collected during any one flight.

However, even where a commercial drone only carries one visible camera, one flight and one set of high definition images can still produce a range of information with multiple end-uses after processing.

The Types of Images Collected

Videos and photographs taken with any camera are georeferenced and are usually high-definition and high-resolution images.

Videos and photographs are recorded onto micro SD cards carried in the camera or in the drone itself.

Photographs can be taken manually as a mix of isolated shots or in a pre-set automated sequence of overlapping photographs for a given area.

Capturing a sequence of overlapping photographs in any format (i.e. visible or invisible light wavelengths) is the most common approach for area-wide inspections. Flight planning software creates an accurate flight path that will ensure the correct overlapping of photographs over the surveyed area. The images are then stitched together and processed post-flight using secure, sophisticated software to create 2D maps and/or 3D models of the areas that have been overflown. The maps are created in natural colours as well as false colours to help highlight variations in reflected *light* for a specific type of information more effectively. These maps are viewable directly and securely online and are also downloadable for use with other computer programmes, or are exportable to overlay on *Google Earth*. They are also GIS ready.

You can view your farm during flights as a live feed, immediately post-flight in a *raw* format or within 24 to 48 hours after processing, depending on the information that is required.

Image Resolution and What this Means to You

The cameras on our commercial drones provide an average image resolution of two to eight centimetres per pixel at 120m above ground level. In other words, each pixel covers two to eight centimetres on the ground. This compares to the image resolution from satellite imagery which is typically meters per pixel! Where the area covered per pixel becomes smaller you gain more accuracy and more valuable information on what amounts to a plant by plant inspection basis.

So, what types of information can you expect to obtain; and, importantly, how can you use the information to improve your farm productivity?

SEEING THE VISIBLE

In this section I summarise the types of information you can collect using normal RGB or visible-spectrum cameras – a what-you-see-is-what-you-get perspective.

It is possible to use software to extract additional, *hidden* information from normal photographs using a variety of conversion formulas and algorithms.

I will explain the extensive value of these image processing options to farmers in the next section, *Converting the Invisible to Visible*.

Information Available from High Quality Videos

At the most basic level of use, normal cameras on commercial drones collect high quality video footage of the areas overflown. Whether you're able to view this video footage in-flight or immediately post-flight provides a considerable amount of valuable information about the physical state of your land, vineyards and orchards from their appearance alone – i.e. without any further processing.

Up to a certain degree, simply collecting video footage allows for real-time crop scouting and immediately replaces the need to drive and walk your fields.

You can instantly pinpoint obvious physical variations in crop growth over a given area that may indicate soil, plant or water problems.

You can observe gaps in your crops due to lightning damage or pests (such as gerbils) or from extreme weather events. You can see if your pivot systems are leak-free or if all the sprinklers are working. You can inspect your boundary fences; or you can locate tears and holes in shade cloth/plant screens, to name but a few examples. I am sure you can imagine other examples where a bird's eye view of your fields, vineyards and orchards would help you enormously.

The video footage can be replayed as many times as may be useful. It can be paused for *screenshot* images or *snapshots* to create a photograph from the footage for any particular feature that caught your attention.

The videos can be easily shared with others should you need second opinions on a problem. And it can provide proof of damage or help with issuing instructions for repairs or further inspections.

Video clips that have been captured for the same target area or subject over a period of time could be edited into a time-lapse video for that period. This enables, for example, a growing season to be compressed into minutes, providing a useful way to visualize and document progress or change over a given area.

And you can always have the videos edited to create marketing material for your business.

However, with videos alone, you are barely tapping the full scope of informative benefits that are provided by recording photographs.

Information Available from High Quality Photographs

Photographs from any camera are arguably more versatile in the information they can provide than video footage, especially where sequences of overlapping photographs are taken over a given area. This is because these sequences of images can be converted into 2D maps, contour maps and 3D models using a variety of online processing (and analytical) software.

Sequences of photographs are best recorded flying autonomously using a pre-planned GPS-guided flight plan and survey grid. This ensures that the correct overlap of each photograph is achieved for accurate stitching and mapping purposes. Trying to achieve the correct overlap manually is difficult, and time-consuming at best.

Once created, a flight plan can be reloaded and repeated accurately as many times as needed.

The set of overlapping photographs are seamlessly stitched together into one image using any one of a number of cloud-based software mapping platforms to create orthomosaic maps.

Established software platforms include *Drone Deploy, Pix 4D, Atlas* and *Precision Hawk*.

The real power of the software mapping platforms lies in the sophisticated analytical tools they provide. These tools convert, interpret and quantify the spectral information captured into crop health data or provide other spatial and quantitative functions for measuring distances, counting plants and calculating volumes.

This extrapolated information is presented as layers which can be toggled on and off when viewing the maps individually, or compared side by side on laptops, tablets or cell phones.

As these maps are both georeferenced and adjusted for the terrain, they can be used to accurately measure distances, areas and volumes for a variety of circumstances that can help you plan more effectively.

This feature is particularly popular for measuring crop damage, or roof damage for insurance claims. Damaged buildings or an area of damaged crops can now be calculated efficiently and with more accuracy than traditional methods. This in turn gives you a better chance of making a more realistic loss claim and securing the claim with less haggling over the extent of a loss. It is also useful for measuring, for example, crown diameters.

In addition to the analytical power of the online software packages, there is an ever-expanding market of custom designed add-on apps that the online software providers are making available on their platforms. They are either in-house apps or have been created by other agribusiness companies, such as *John Deere*, for use with their management software such as the *John Deere Operational Centre*.

The apps provide additional analytical tools, interpretation tools, conversion tools or enable data to be used with other systems, including, for example, high-tech machinery with variable rate application capabilities for the precision application of nutrients, pesticides, fertilisers or water irrigation. In other words, the apps are providing more opportunities to extrapolate multiple information sources from a single set of photographs recorded by normal visible-spectrum cameras alone and convert this information into functional data or instructions.

The types, number and scope of the apps are evolving all the time in response to farmers' needs, as are the online software platforms. For example, it is already possible to convert normal visible footage into multispectral image approximations for crop health assessments.

Similar algorithms are in the pipeline for converting visual footage into alternative and insightful data sets. More functions are being added which are significantly enhancing the versatility and value of the data being collected by drones with standard and specialist cameras.

Finally, images can also be converted into 2D contour or elevation maps and 3D maps using the same online software platforms.

The contour maps are created as *Digital Surface Models (DSM)* and illustrate the ground elevations and slope gradients of the area that has been photographed, using false colours to better highlight the differences.

Once generated, you can use this information to repair and maximise the effectiveness of the natural drainage and field irrigation systems after a season, or help identify new dam locations.

The 3D maps require a different approach during flight to capture enough photographs, but they are useful visualisation tools for calculating volumes of stockpiles, illustrating landform and estimating the heights of crops. Both the contour and 3D maps are not as accurate as the maps generated by surveyors using multiple ground stations, or by commercial drone systems equipped with LiDAR systems but, again, they are still significantly more available for making informed estimations and more easily obtained compared to traditional methods.

CONVERTING THE INVISIBLE TO VISIBLE

This section covers the range of information that can be gleaned where commercial drones carry specialist cameras that are sensitive to non-visible wavelengths, including the multispectral and thermal imaging cameras.

This is where the advantages of drone-captured information really shine and give you the substantial edge that you want over the challenges you currently face in determining the state of your soils and crops, compared to traditional methods.

World-wide, this is where traditional agriculture is being turned on its head and drone supported 'smarter' precision farming is providing opportunities for you to achieve more control over your operations and productivity.

Why Drones Carrying Multispectral Cameras are Powerful

Just to give a short background on how the specialist multi-spectral cameras work, all objects reflect light.

Objects are the colour we see because they reflect the most amount of light in the specific visible wavelengths we can detect. In other words, a healthy plant appears green because it reflects more light in the visible green wavelengths than other visible wavelengths.

Plants also radiate at other wavelengths, typically in infrared spectral bands. This creates a *spectral signature* which varies depending on the plant species as well as on the health of the plant or the stress it is experiencing at the time.

Significantly, less than 30% of a plant's reflectance falls within the visible light-spectrum with the largest proportion (whether the plant is healthy or stressed) falling within the infrared spectral bands.

Plants look green and healthy even when they are under stress, because the deteriorating condition is not initially detectable in the visible spectrum (i.e. green wavelength). However, the warning signals would already be apparent at near infrared wavelengths. It therefore makes sense to monitor your crops at infrared wavelengths, if you want to understand what is really happening and want to be forewarned of a pending problem.

In contrast, by the time there is a noticeable change in the colour of the crop, the opportunity to remedy or rescue the crop may have been lost.

As the composition of reflected wavelengths or spectral signatures are plant specific and predictable, it is possible to correlate the resulting spectral patterns with the condition of the crop and, indirectly, the condition of the soil they are in.

This is achieved using plant indices which are based on known relationships between selected wavelengths or bandwidths that correlate with a specific crop or soil conditions directly or by association. A multitude of variables for any given field, crop vineyard or orchard are exhibited through the reflected invisible wavelengths, including soil fertility, plant health, biomass and chlorophyll levels. [13]

As different crops have different spectral signatures it is possible to identify threats from weeds from the same set of images. This is a major benefit for any farm.

It follows, multispectral cameras are designed to capture the bandwidths that best match the crops' reflectance from which the condition and that of the surrounding soil can be determined.

There are several multispectral cameras on the market.

These cameras take multiple photographs of the same point in different bandwidths, depending on the number of visible and infrared bandwidths the camera has been designed to accommodate.

The camera we currently use, the *MicaSense RedEdge*, has five lenses and each lens is sensitive and dedicated to one of the following bandwidths, red, green, blue, red edge and near infrared. The resolution of the *MicaSense RedEdge* is eight centimetres per pixel at 120m above ground level.

Why Drones Carrying Thermal Imaging Cameras are Powerful

In contrast, the thermal imaging cameras detect heat signatures and heat differences. They work on the basis that all objects above absolute zero (minus 273º Celsius) will radiate heat or thermal energy.

Thermal imaging cameras can detect heat and differences in temperature to potentially very high sensitivities. These heat vision cameras are capable of detecting and displaying very small differences in temperatures. Cameras with high sensitivities coupled with high resolution tend to produce better quality images. This in turns gives more flexibility in the use of the camera; the close inspection of images; adjusting for contrast and noise; and problem diagnosing potential. [14]

The footage is displayed in a black to white grey scale or brighter rainbow colour options depending on the model of the camera and/or the software or supporting apps used.

Thermal cameras are trickier to use than the other cameras because of the errors that can be introduced by flying too fast or not allowing for current weather conditions which affect the temperature profiles that will be identified by the camera.

Some thermal imaging cameras are also more selective and specific to a given range of operations and therefore less flexible in their applications.

Where carried by drones, thermal imaging cameras are particularly beneficial for water management and disease detection through leaf surface or canopy temperature differences. For example, crops that are water stressed tend to heat up because they close their pores to reduce further water loss. Similarly, disease can also produce the same effect. This means temperature differences can be used to guide irrigation schedules and target crop inspections.

I have more experience of using the multispectral camera for agricultural uses.

The use of thermal imaging cameras and other sensors with commercial drones in agricultural applications are still in their infancy and not yet widespread.

However, there is already a lot of excitement and anticipation about the immense value these cameras and sensors will bring to farmers and growers. *'Thermal is going to be huge and there are some other new sensors being developed ... this is new technology that will change the game of agriculture.'* [15]

Information from Multispectral Cameras

The multispectral cameras capture only capture photographs.

Similar to using a normal visible light-sensitive camera, single photographs or a sequence of overlapping photographs of an area are taken using the multispectral cameras. The subsequent processing and analysing of the high quality, geo-referenced and calibrated orthomosaic maps with multiple plant indices and digital surface layers substantially leverages the value of the information that has been collected.

The geotagged, colour composite layers and plant index maps reveal specific crop and soil conditions.

For example, the *MicaSense RedEdge* images provide the following layers and maps through the *Pix4D* and *Atlas* software: [16] [17]

- The *Normalised Difference Vegetation Index (NDVI)* has long been in use with images collected by satellite or manned aircraft and it the most commonly used index. An NDVI image represents the amount of biomass, scaled from zero (no leaf) to one (multiple layers of vegetation). Biomass difference can indicate differences in plant vigour or leafiness, differences in soil water availability; conditional foliage nutrient content (including nitrogen); and yield potential. NDVI tends to be more useful from early to mid-season.

- The *Chlorophyll Map* shows chlorotic stress and it is used for identifying vigorous healthy crops; estimating chlorophyll content and conditionally estimating nitrogen. This map is most powerful in combination with NDVI maps as it provides a better measure of overall plant health. It is only available with a multispectral camera that is sensitive to the red edge bandwidth.
- The *Normalised Difference Red Edge (NDRE)* is only available with cameras that capture the red edge band width, such as the *MicaSense RedEdge* camera.
  The red edge bandwidth is very sensitive to plant stress and NDRE maps show: leaf chlorophyll content; plant vigour; stress detection; fertiliser demand; and nitrogen uptake.
  In comparison with NDVI, the NDRE index provides more information on plant nutrients. Some suggest NDRE is best used mid to late season. However, comparing NDRE and NDVI patterns can help show where crop lag may be affected by low nitrogen or other causes.
- *Colour Infrared composite (CIR)* is, as the name suggests, a composite rather than an index. It emphasises the reflection of the near infrared band widths to help assess plant health; identify water bodies; variability in soil moisture; and soil composition.
- The *Optimized Self-Adjusted Vegetation Index (OSAVI)* indicates canopy density and plant vigour.
- The *Digital Surface Model (DSM)* is used for estimating relative crop volumes, crop heights, identifying surface properties and modelling water flow and surface accumulation.

This is not the full range of indices and composites that are available, but just a selection of the more popular options used in agricultural applications.

The final selection and use of these and other composites and indices will depend on the timing of the survey, the season, the time of day, the types of crops being assessed and how the crops have been planted. This is because reflectance sensitivities within the algorithms used to create the composites and indices may favour, for example, early growth or high canopy density.

As with all the commonly used vegetation indices, the images are not wholly prescriptive and, for example, do not pinpoint the exact type of disease or pests attacking a given plant or crop. Crop inspections must still take place. However, with subsequent targeted crop inspections, the plant health specific indices still provide a very powerful time- and crop-saving tool.

Repeating flights and data collection over the same areas over several seasons highlight trends and traits that help you make more informed assessments of the causes of stresses and conditions observed.

Some vegetation-specific indices are already available, for example, *Precision Hawk* has wheat specific indices for assessing the potassium (K), nitrogen (N) and sulphur (S) content in wheat. It is inevitable that within a short space of time image processing will soon become more prescriptive and therefore become more immediately valuable to farming.

There is no stopping us from using multiple software platforms to extract more *invisible* information from the photographs we collect. Exporting the same set of visible or invisible spectrum photographs into other software platforms, such as *Precision Hawk*, adds more value to the visible and multispectral data with access to the additional composites, indices, apps and automatic report generating features, including, for example:

- Canopy or vegetation cover;
- Stand counts and plant counting;
- Individual tree crown and health and selective vegetation indexes, depending on the camera used.

As with visible spectrum photographs, accurate measurements of the physical areas of diseased crops, stressed crops, or healthy crops can be calculated from the images.

It is noteworthy that photographs captured by standard visible cameras can be used to create vegetation index maps using selected software platforms, such as *Drone Deploy* and *Precision Hawk*, for basic crop health assessment.

These platforms use the *Visible Atmospherically Resistant Index (VARI)* and *Visible NDVI*. They are an alternative to the true NDVI where specialist cameras are not available or not in use.

The resulting maps or layers give an indication of crop stress but VARI or Visible NDVI cannot replace the scope of information available via multispectral cameras. Even *normal* cameras fitted with a near infrared filter are second best to the true multispectral camera, though this option is better than using the normal camera alone.

The VARI gives a good insight of general crop health, but only at the time the photographs were taken. Ground-truthing is essential before any actions are taken.

These maps will not be reliable for monitoring changing crop conditions over multiple time periods during a growing season because of errors introduced by differing amounts of sunshine or cloud cover during each set of photographs. This error does not occur with photographs taken using multispectral cameras, as the images are calibrated. [18]

Once the processing has been completed, we can use the software to convert the maps and layers into a selection of formats, including KMZ files for *Google Earth*, or conversion to formats for export with other farm management packages *'to create prescriptions for variable rate application'* pdf documents or for GIS applications. [19]

Information from Thermal Imaging Cameras

Thermal imaging cameras can capture both video footage and photographs. The video footage can be viewed in-flight or post-flight, depending on the model and make of thermal imaging camera being used.

When used for agriculture applications, thermal imaging cameras pick up variations in temperature that are not only an indicator of a plant's stress, but also its metabolism, fruit maturity and yield.

These variations also indicate the presence or absence of water in the soil. As with visible spectrum video footage, there is an element of real-time field or crop scouting management that can be achieved with thermal imaging video footage.

As an aside, thermal imaging cameras are also proving very useful for livestock farming, farm security surveillance and building maintenance inspections, whether being carried on commercial drones, by vehicles, or as handheld devices.

TO SUMMARISE

Going back to the questions I listed at the beginning of this chapter, you have now seen the range of information that is available via commercial drone-powered surveying, the cameras they carry and the powerful software used before, during and after each flight.

Collectively, and whether using visible or multispectral or thermal imaging cameras, there is a huge range of information that is now available to you, including, at a minimum:

- Soil nutrient conditions, as indicated by plant nutrients;
- Surface elevations or topography;

- Plant and structure damage and loss estimates;
- Plant counts, canopy cover and leaf areas;
- Crown diameters and crop heights;
- Flowering estimates;
- Plant vigour and plant health assessments throughout all growth stages;
- Water availability to plants;
- Water in soils;
- Presence of weeds, diseases and pests;
- Relative biomass estimates;
- Area and volumetric data;
- Growing trends and traits (with repetition);
- Asset inspections.

It is important to highlight that the mapping results are not wholly prescriptive – just yet.

At the moment, the imagery will show any area of crop stress, but targeted crop scouting will be needed to confirm the nature of the stress, e.g., a rust fungus in wheat or extensive weed growth and so on. Drone surveying results are best used in conjunction with ground-truthing to confirm the cause of the variations identified and the full and right scope of actions required. Time has been saved, more problems identified than might have been missed, and problems captured early.

The benefits of having easily available, good quality information about your fields, vineyards or orchards is the foundation for resolving both the second and third core problems you may experience.

Next, we shall see how being better informed about your soils and crops releases you from time-consuming, non-productive activities.

Part 4
Spending More Time
On Productive Tasks

Part 4: Spending More Time On Productive Tasks

From our experience, a key problem for you will be spending too much of your time on mundane tasks at the expense of productive tasks. This is not necessarily through any fault of your own, but rather a by-product of operating over large areas alone. Precision farming will enable you to spend more, good quality productive time on your farms, vineyards or orchards.

How would you rather spend that time? What would you rather be doing when making time-consuming plant counts, walking or driving round your fields, vineyards or orchards for plant inspections or checking irrigation systems and pivots? How would you rather have your teams spending their time?

I would suggest you would much rather spend time being more productive responding to the information collected about your soils, crops or equipment and structures to improve your yields, rather than trying to collect it.

Drones provide the rationale and create the opportunity for you - and your teams - to save time otherwise spent on less effective mundane, time-consuming tasks, especially crop scouting and infrastructure inspections. Even allowing for the fact we would complete the inspections for you, agricultural drone surveying still gives you additional opportunities to use your time even more productively.

There are four features from using our commercial drone services that release you from mundane tasks as well as give you a chance to gain an advantage over time and react more timeously.

TIME SAVING TECHNOLOGY

Compared to manned aircraft and satellites, commercial drones can be flown on demand and collect real-time, high quality and high definition information rapidly and efficiently.

In our case, we can be deployed and re-deployed quickly and as often as needed, and both throughout the growing season and in-between seasons.

This flexibility is especially advantageous compared to manned aircraft where, for example, unstable weather can delay or postpone flights or render satellite information useless when there is cloud cover.

By contrast, agricultural drone surveying can be deployed quickly once an appropriate window in the weather opens up.

Drones are reliable aircraft for data collection.

As a practice, we take two drones to site, so we have one as a back-up in the event of an unlikely technical problem with the primary drone. This is rarely an option with satellites or manned aircraft operations!

Pre-Set GPS Flight Planning

We fly drones manually or autonomously using a GPS pre-set flight plan. Pre-set flight plans are best used for creating accurate survey grids of the field, vineyard or orchard we are investigating.

Importantly, these accurate flight plans ensure there is a correct amount of forward and side overlap of photographs within the defined area so that there are no gaps in the processed images. The pre-set survey flight plans for autonomous flying are straightforward to prepare by a trained operator, and, importantly, can be stored and then reused as many times as necessary without any further inputs. This helps save time for redeployment. This feature can be critical if you need to repeat accurate surveys over time to build up a data bank for comparative analysis and longer-term performance assessments.

Importantly, we can collect information throughout the growing season without interrupting your other farming activities.

Drones fly directly to an area of interest at speeds of up to 72kph for a *DJI Phantom 4 Pro* series or 93kph for the *DJI Inspire 1* if we applied full throttle, though this is not really necessary.

We can fly over obstacles or quickly reach areas that are inaccessible on foot or via vehicles or would require lengthy detours. The average range of a multi-rotor or a fixed wing drone is between two – seven kilometres from the pilot. However, rules and regulations exist which set conditions for the distances that may actually be flown.

Visual Line of Sight (VLOS)

In South Africa, recreational drone uses are limited to *visual line of sight (VLOS)*, with no legal options to operate *beyond visual line of sight (BVLOS)*.

This means *recreational drone* users are restricted to flying a maximum horizontal distance of 500m and a maximum vertical height of 120m above ground level from the operator.

Commercial operators have more operating freedom, which depends on the types of operations that they have been rated and licensed by the civil aviation authorities to undertake and can continue to prove they do so safely over time.

Even within these legal flight distance restrictions, the distances are covered quickly and soil or crop intelligence collected is still immensely valuable to farmers. Drones cover and record information over large areas in a single flight. This means better and more information can be gathered more quickly than achieved without drones.

Fixed Wing or Rotors?

The time in-flight and the physical area covered depends on the type of drone used (fixed wing or multi-rotor).

As you would expect, there are pros and cons using a fixed wing drone or a multi-rotor drone. Fixed wing drones tend to cover larger areas as they have a longer battery life, and hence flight time. However, there is a versatility and accuracy of positioning with multi-rotor drones which offsets the disadvantage of shorter flight times of multi-rotors, which we prefer.

We fly multi-rotor quadcopters, including the *DJI Phantom 4 Pro* and the *DJI Inspire 1*.

They traverse an area of roughly 60 hectares in an overlapping grid pattern in 25-30 minutes at a height of 120m above ground level.

In other words, they can effectively map roughly three hectares per minute.

The *DJI Inspire 1* has a slightly shorter time in the air than the *DJI Phantom 4 Pro* but its advantages lie with its ability to carry different cameras, including our *MicaSense RedEdge* multispectral camera.

Alternatively, a fixed wing drone, such as the *Sensefly eBee SQ*, covers approximately 200 hectares in a single 55-minute flight 120m above ground level. [20]

TIME SAVING ANALYTICS

Not only can commercial drones collect high-quality, high definition information more quickly, but the information can be processed and organized into actionable information much more quickly than traditional methods.

We have seen in Part 3, commercial drones provide a versatile and flexible vehicle for carrying different types of cameras which quickly capture valuable, high quality, real time information about the physical and/or biological conditions of soils, plants and built structures in place.

Additionally, a vast amount of information is captured by the cameras directly and through the processing software, which converts the information into geotagged interpretations of the videos and photographs. One flight produces multiple types of information, which means you can use the data from one flight to, for example, check for water stress, the presence of disease or pests, review soil conditions, plan natural drainage repairs, confirm irrigation system blockages or observe lightning or hail or storm damage and/or calculate physical areas of crop loss for insurance claims.

Faster Turnaround Times on Data

The turnaround time we experience with the software platforms from flight to interpretable data ranges from a matter of minutes to approximately two days.

Once rendered the information can be easily shared and put to use more quickly than previously possible with traditional data collection.

It is worth bearing in mind the turnaround time for rendered *stitched* images and multiple layers has decreased rapidly and online software platforms are driving towards reducing this time further. This enables you to respond more quickly to any problems the data has identified. This is a feature the software companies are competing to get right in order to enable truly real-time data gathering, assessment and decision making.

It is only a matter of time before the full scope of detailed high-resolution mapping information described in Part 3 will be available in real-time and while the aircraft is in the air. [21]

Better Reports Mean More Timely and Relevant Information

Other advantageous features available to us with some of the software platforms includes the ability to immediately convert data into pre-formatted illustrated reports with automatic annotations, positioning data and date stamps; all at the click of a mouse. This includes, for example, calculating plant counts and yields or estimating economic loss per field.

The high-quality, high resolution imagery means you can now effectively manage plants rather than fields at a time, which offers more efficiency through, for example, precise inspections and variable rate applications.

TIME SAVING DECISION-MAKING

The turnaround time from flight to decision making can now be measured in minutes to a couple of days.

This means, we are dramatically collapsing information-to-action timeframes.

This, coupled with your having more information about problems that may have otherwise been missed, offers enormous opportunities for you to concentrate your actions on resolving smaller problems before they worsen into major, complicated and time-consuming problems.

This compares to the situation described in Part 1 where you spend a lot of time trying to cover your fields, orchards etc to inspect your crops and even then, problem identification can be a hit or miss affair and down more to luck than by design.

In addition to being able to save time by being able to react sooner to a problem, you subsequently gain more time to direct your activities elsewhere on the farm at more opportune times because you are less tied up with responding to otherwise major problems.

The digital information we generate can be stored, archived and shared easily on any mobile device and reviewed as many times as needed and as instructed or controlled by you. This gives greater flexibility for you when communicating and resolving problems with your teams or specialists without, for example, more time lost due to additional inspections and travel.

As a side note, the software platforms we use are subject to strict privacy and data protection laws. We will also respect your privacy with all hard and soft copy information we generate in connection with your farm. Any subsequent sharing of information will be under your control.

TIME LAPSE

With repetition and viewing data sets side by side, you have a more holistic view of your fields, soils, crops vineyards and orchards. With easier accessibility to this data, you can start to compare and assess performance over a single season or across multiple seasons more quickly.

Again, I cannot overemphasize just how significant this benefit is to your farm.

'With a greater understanding of what happened, exactly where it happened, and at what point in time, you can make smarter crop management decisions to maximise next year's yield.' [22]

TO SUMMARISE

Our agricultural drone services and the information they collect are available on demand and reliable.

Vast amounts of information can be collected efficiently in real-time or near-real time both directly or through subsequent processing and interpretation.

Time-lapse reviews and comparisons of the data sets collected over single or multiple seasons offer more opportunities for identifying trends and traits that can help you save time with future growing seasons. Processed information is available to you in a variety of exportable digital formats for quicker interpretation, dissemination and action as you may need.

Currently, commercial drone services will not completely eliminate the need for you or your teams to physically inspect your plants, soils or structures, but it does mean your time can become focused and productive because you spend less time traversing entire vineyards, fields or orchards and being directed straight to problem areas timeously.

Once you are more reliably informed about what is happening on your farm and you have more time to be productive, it follows that you can prioritise, strategise and manage your operations far more effectively and reliably.

Suddenly, you can plan with a degree of certainty, react more quickly and have more time for creating better yields.

Part 5 explains how farmers truly become smarter

farmers as a result of having the scope and types of information that are now available to them more quickly and timeously.

Part 5:
Creating Better Management Strategies

Part 5:
Creating Better Management Strategies

Information and the ability to react at the right time in the right way are the cornerstones of better, effective management of any business. The same applies to modern farming.

We have seen in Part 3 the unprecedented level and detail of information now available to you through commercial drones, the cameras they carry and the online software platforms.

We have also seen in Part 4 how the acquisition of this information is releasing you from traditionally time-consuming tasks as well as giving you time to react to problems that may have gone unnoticed or been discovered too late to save.

Part 5 covers how the combination of these two components of drone supported agriculture leads to more precise, efficient and sustainable management practices or smart farming.

So, how do agricultural drone services give you the edge over the traditional risks and challenges you have faced thus far?

And how are they transforming farming practices?

SMARTER RISK MANAGEMENT

Critically, you can now identify a far broader range of problems with your soils, crops or equipment that currently are overlooked, missed or go undiscovered or undetected. These problems may be overlooked or go undetected because of limitations with what the human eye can physically see as well as traditional data gathering options.

Commercial drone surveying provides such precise, high grade information that you now have an early warning system on the risks you usually face with disease, pests, weeds, extreme weather, human error or simply troublesome equipment. Not only does commercial drone surveying provide this information but it can do so much earlier. This allows you to respond more quickly, effectively and typically with fewer resources and/or lower costs.

Organising responses to problem areas are more accurate. Teams can be directed almost on a plant by plant basis to within meters of the problem area; the area requiring action can be accurately quantified for repair options; and, where required, information can be easily converted for the *immediate* variable rate applications of chemicals, fertilizers or water at the affected areas using specialist equipment.

Accurate assessments can now be completed using commercial drone surveys, including plant counts, crop losses and estimated yields, when you are experiencing losses. This means insurance claims can be more pain-free because they can be resolved more efficiently and pay-outs received with less delay.

Where you are relying on other service providers to construct new fences, erect new shade cloth or transplant plants, you can have their progress or completed work surveyed easily and quickly thereby protecting yourself from accepting or paying for poor quality work or substandard service delivery.

'These advantages help farmers catch problems faster and react more quickly, which can save thousands of dollars in crop losses per field.'[23]

From the perspective of risk management alone, professional drone surveys are increasing the productivity and profitability of your farm by reducing your vulnerability to environmental risks and man-made issues and the losses that may otherwise occur.

SMARTER RESOURCE MANAGEMENT

Professional drone services provide detailed information that give you precise control over your use of resources, where a resource constitutes seeds, crops, chemicals, fertilisers, equipment, energy (fuel and electricity), water, labour and/or time. Consequently, you are gaining control of your expenses by more effective management of your resources than otherwise occurs in traditional farming.

This invaluable information is readily converted into actionable data which means you can: reduce over-applications and wastage of chemicals, fertilisers and water on your plants and soils; use less fuel in your vehicles and machinery; spend less time responding to issues; optimize your natural drainage systems to save electricity demands for pumping water; monitor dam or river levels; and capture water losses due to faulty irrigation systems, to name but a few examples.

In other words, resources are used more efficiently and conservatively but with a higher and targeted productivity.

Where you choose, data can be easily shared and distributed to teams, specialists and other parties enabling effective collaboration for assessing problems;

promoting more informed decision making; ensuring clear understanding of instructions for action; pinpointing target areas for responses; and generating accurate qualification of areas and dimensions.

The information provided by professional drone aerial surveys increases a farm's productivity and profitability by achieving more with fewer but precisely directed resources.

It follows the environmental performance of your operation will also directly improve.

SMARTER BUDGET MANAGEMENT

Professional drone surveying and the information produced means you can make game-changing improvements to your operations, such as implementing smarter risk management and smarter resource management. This typically means you can grow more with less expenditure.

Furthermore, not only are you able to produce your grains, grapes, fruits and nuts for less cost, but you are able to generate better quality products due to the improvements with plant management practices. These same improvements raise the value of your harvest as well.

Being more informed about the quality of your yields throughout the growing season means you can forecast your income and spending more accurately (see below).

Quicker pay outs for insurance claims, facilitated by accurate loss assessment maps, can also help maintain funds and cashflow.

Professional drone surveys are increasing a farm's productivity and profitability by reducing your vulnerability to unpredictable costs, reducing your costs, at the same time as increasing the productive use of available budgets and income generation potential.

You can also gain a greater flexibility to re-invest in the farm and weather the unexpected than you may be experiencing with traditional farming practices.

SMARTER YIELD AND HARVESTING MANAGEMENT

The resolution of the information collected from commercial drone surveys and rendered using the software platforms enables you to manage on a plant by plant basis as opposed to a more random and field by field basis. This means you can streamline and optimize your activities and resources more accurately to produce better quality and higher yields.

Information collected using commercial drone surveys is now giving you the ability to calculate your yields quickly and with more accuracy than previously available, thanks to automatically generated plant counts. It also assists with pinpointing the best time to harvest. This means you will be more accurate with your budget forecasting as well as organizing your harvesting resources such as hiring help, machinery and transport.

Drone surveys, using specialist cameras and software platforms, are increasing a farm's productivity and profitability by improving yields (quality and quantity) through effective risk management, precise resource use and more accurate forecasting.

SMARTER MANAGEMENT FOR YOUR FUTURE

'Long-term, the data generated by drones help farmers gain a more accurate and detailed picture of how their crops are reacting to their management strategies, which can lead to more effective use of limited resources.' [24]

You gain more opportunities to plan more accurately for the next season and more seasons beyond. This follows from the accumulation of information for the same areas over a single or multiple and consecutive seasons.

Sequential data obtained via agricultural drone surveying systems can be easily compared to reveal trends in the behaviour of soils and plants to specific conditions, treatments and farming practices.

Analysis of these trends can help guide your future activities to avoid recurring problems or repeat and build on successful traits.

The value of these comparisons can be enhanced by adding data collected by traditional methods, such as detailed soil sampling to corroborate the drone supplied data and obtain more confidence in predicting the cause of problems or for defining more accurate variable rate prescription for nitrogen, pesticides and other applications.

Some of the software platforms allow maps to be imported from other farming software programmes (e.g. the *John Deere Operations Centre*) to be able to make comparison with harvest, yield, variety and spraying maps. This means you can *'look deeper into problems areas and find better, effective responses.'* [25]

You can also gain more planning and management control for future seasons through the benefits of smarter resource, budget and yield management.

This means over a longer term, you can gain more financial resilience and flexibility to expand your farming operations for better profitability in a more sustainable way. For example, you can hold on to your produce for better prices rather than sell immediately where prices are lower because of cash flow or budget pressures.

Drone-supplied information can help you manage and use the time between growing seasons more efficiently to repair and prepare the drainage systems for the fields, vineyards or orchard more effectively.

Drones are increasing a farm's productivity and profitability by increasing the accuracy of forecasting for the management of risks, resources, budgets and yields for the next season. They are also helping farms improve their long-term operational and environmental viability through more efficient resource use.

TO SUMMARISE

At the end of Part 1, I asked you what your burning question might be, the ultimate *'How do I….'?* question which, if answered would give you what you wanted for your farm, vineyard or orchards. I know the majority of farmers would ask *'How do I achieve higher yields and better profitability?'* and I promised you I had the answer to this burning question.

I have no doubt professional agricultural drone services and the sustainable land management practices they directly promote are the answer for agribusinesses.

I have used this book to present why I am absolutely convinced this is the right the answer.

So, the next question is, *'How do I do this?'*

Part 6:
The 5-Step Precision Farming Blueprint To Smarter Farming

Part 6:
The 5-Step Precision Farming Blueprint To Smarter Farming

We implement a 5-step precision farming blueprint to help you achieve better yields and increased profitability. The more this process is repeated, the more valuable the information collected becomes and the more improvements you will see with your productivity, agribusiness management and ultimately profitability over time.

I have outlined the key actions we take during each of the five steps. For the sake of an easier read, and for our international readers, I have not gone into the nitty-gritty of the mandatory processes and procedures we are required to perform in terms of *South African Civil Aviation Authority* approved operations manual.

Most countries are now moving towards the establishment of rules and regulations for facilitating safe recreational and commercial drone use within their airspace.

South Africa takes pride in being one of the first countries to establish regulations and has one of the strictest frameworks in place for all drone users but particularly for commercial operators.

We have adopted a sequence of procedures and checklists we must follow and complete for every mission – from appointment to completion.

The documents we generate for each mission are stored for annual inspection by the SACAA as well as our own internal inspections. Suffice to say, in South Africa, we take our legal obligations under the *South African Civil Aviation Regulations* very seriously.

Never mind that failure to do so would be highly unprofessional and could lead to legal consequences and grounding; we do so because we acknowledge the need to ensure safe operations take place above your property and within an airspace that is shared with other aircraft. The rules and regulations are also in place to protect the public's safety and privacy and ensure a proper and professional service delivery takes place.

In South Africa, all commercial use of drones must be performed by licensed pilots and licensed operating companies. In one form or another, all countries are moving in this direction to ensure safe use of drones world-wide.

With this in mind, this is our *5-Step Precision Farming Blueprint:*

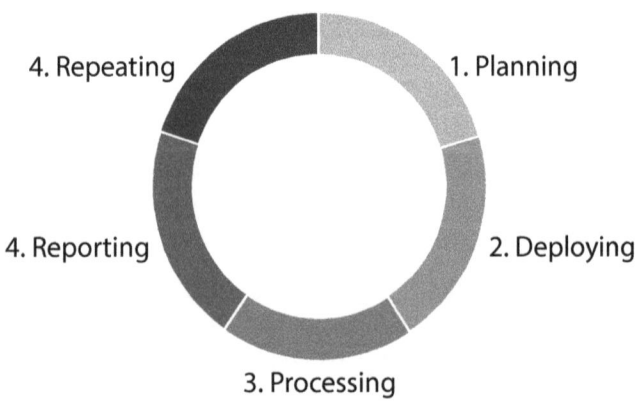

STEP 1: PLANNING

Step 1 is essential to understanding your needs and for meeting our professional service and legal obligations. It comprises a mixture of activities that take place either on-site or remotely.

First, we confirm what you want to accomplish

This is essential for us to confirm the types of information we need to collect, select the right equipment for collecting the information needed and complete the pre-flight preparations including, for example, site safety assessments, pre-flight equipment checks, flight planning and weather checks. We prefer to meet with you directly, however this can be completed remotely and provided we have the location to review via *Google Earth* for our flight planning purposes.

Secondly, we plan the actual flying mission

It is particularly important we have a clear understanding of the area we are flying over as well as the surrounding area so we can confirm if there are physical obstructions that could, for example: prevent the pilot maintaining radio or visual line of sight with the drone; restrict views from our cameras; prevent us from flying over the entire target area because of airspace restrictions; produce collision risks (e.g. with overhead wires or overhanging trees); contain other legal restrictions that would limit the effectiveness of all or part of the mission.

It is important we clarify any difficulties, impediments or risks as soon as possible so that we can determine alternative and safe approaches for collecting the information you need.

This is usually done at our office with some on-site

knowledge of your farm or survey area so we can plan our take off/landing sites and create our survey grids for automated flying.

We also select the drone, the camera, the software and then create a flight plan that will best suit the survey we will be performing that meets your requirements.

Finally, we check the long-term weather forecasts to find the best periods for flying

The best times to fly and collect data are typically between 1000 - 1400 hrs for multispectral vegetation mapping to minimise shadows. Wind speeds should be anywhere from calm to 12m/s and there should be no risk of rain, thunderstorms, fog or mist.

We also find it is better to fly in the morning to avoid afternoon changes to wind speeds. It is also better to fly with either clear skies or consistent overcast skies where data is to be collected to assess the condition of vegetation. If necessary, we'll come up the day before to make sure all is in place for a good day's flying and maximum data collection.

STEP 2: DEPLOYING

When we deploy we are on-site and flying our drone over the specified area, using the camera appropriate to the information you need to solve your specific problems and/or using the right software support as defined by the mission at hand. This is where we first get to observe what is happening on your farm.

It is always exciting to immediately gain a new perspective on your situation and start to see the opportunities for solving some of your problems straight away.

During this time, we perform as many manual or automated flights as necessary to capture the information required or cover the physical area of interest. We want to ensure we gather enough video footage, single photographs or a sequence of photographs using normal or specialist cameras.

We begin by setting up our location for launching (and landing) our drone safely and where we can continually observe our drone during flight.

Once we have the drone in the air we either complete the mission by flying manually or we will fly on autopilot using the pre-determined flight path and survey grid for accurate filming or photography.

Video and/or photographs captured manually or at pre-programmed intervals are stored directly on a microSD card carried by the drone.

If you choose, you can view our flights directly via a livestreaming option or by watching a second screen with the camera operator. The availability of a second screen depends on both the camera and model of drone we are using at the time.

We capture images from any height from several meters to up to 120m (400 feet) above ground and up to 500m away from the pilot (for the time being). [26]

Depending on which model of drone we are flying, we have multiple batteries that currently give us a total of up to 90 minutes flying time from one combined charge. We recharge our batteries on site and generally tend to rotate each battery for recharging as soon as they are depleted and cooled down where an extended mission requires. Battery life, and therefore flying time, varies depending on the weather conditions, temperature, altitude and operating conditions. In other words, should it be windy, cold and we're flying at altitude then the battery life and flying time will be reduced.

This is one of the reasons why we have multiple batteries so we can offset any reduction of flying time due to environmental conditions.

Once the drone has landed, powered down and we've completed our post flight procedures, we transfer a copy of the data on the microSD card to a laptop or another device as a backup.

Wherever possible, we usually check the data collected with you on site to make sure that we've captured the correct areas or have sufficient overlap of our survey area to create good quality orthomosaic maps.

If necessary, we will adjust a grid survey and repeat the flight if we feel our first run has not collected enough information or the right information or it becomes obvious we need to expand our survey area to help solve a problem.

STEP 3: PROCESSING

Step 3 is where we leverage the full value of our drone survey services for you via the secure online processing software platforms we choose to use or by simply reviewing the video footage or single photographs immediately upon landing.

This is the step where the raw footage we have collected is quickly converted into information about the actual state of the farm, vineyard or orchard and transformed into actionable data for you to use.

I have already covered the image processing process in detail in Part 3 in terms of the types of information that we extract and generate from the processed imagery.

Processing can take place at the farm (depending on internet connectivity and speeds) or back at our offices.

The turnaround time is anywhere between minutes or

24 – 48 hours for the return of the processed information from the online software platforms, depending on the information and format required and its forthcoming application.

STEP 4: REPORTING

Reporting on the results of our activities can take any number of forms depending on what you wish to accomplish and the types of information we collected.

We supply printed and/or electronic representations of the information that has been collected.

This may take the form of: raw or edited video footage and photographs from basic reconnaissance; written reports comprising annotated maps, interpretations and recommendations for crop scouting or damage inspections; private hyperlinks to the online software platforms we use for the image processing so that you can review the 2D maps and 3D models only; or extracted files for your own use and integration with other programmes.

STEP 5: REPEATING

'By correlating the data obtained from previous seasons with the current data and by comparing the results from the same developmental stage from one season to another, the farmer will obtain a database enabling him to refine his yield estimate even further.' [27]

Step 5 is often overlooked, but it is critical if you want to optimise short, mid and long-term improvements to your management practices, yields and ultimately profitability at your farm, vineyard or orchard.

As I have covered in Part 3, collecting multiple data sets for the same areas over a single season or, better yet, multiple seasons, reveals more valuable, time-based management information.

Similarly, capturing information on the same day or week each year is also very valuable as it gives an idea not only of crop variations but also seasonal variations.

Depending on the plants you intend to grow, we would generally suggest the following programme of drone flights during any given season. These are some of the best phases in the season where the results of the flights can be leveraged for optimum plant management:

- Soil mapping before planting commences;
- As soon as crops begin to emerge to assess germination;
- Throughout the growing season up to harvesting;
- Immediate pre-harvest to assess ripening;
- Immediately after harvesting.

If the flights are repeated over the same areas over multiple seasons, then you can compare and measure your performance and make informed management decisions for the next season.

It is also especially useful to repeat flights over consistently poor performance areas.

The serial footage may help identify underlying problems that are causing the poor performance – e.g. poor drainage – or may even identify the presence of consistently poor yield areas in the first place.

Repetition is not always relevant. One-off flights are valuable in their own right, e.g. after extreme events such as severe thunderstorms and fires, to assess the plant and structural damage, make plans for repairs, estimate losses and prepare insurance claims.

This being said, we usually find that once someone has seen a drone in action, they appreciate the greater opportunities for use.

TO SUMMARISE

Commercial drone surveys are structured and accountable. In essence, we work with you to determine your goals so we can select the best data sets to acquire, using the right drone, cameras(s) and software applications. In other words, we do not arrive, fly, photograph, hand over a micro SD card and leave.

As I have emphasised throughout this book, commercial drone use is regulated in many countries for ensuring a professional and safe service for clients.

This means professional services are based on procedures and processes that have been approved (and continuously audited) by civil aviation authorities. Operators are required to prove they have complied with their approved procedures and processes to ensure they continue to remain licensed and operational.

This means you should receive information that is relevant and meaningful to your agribusiness.

If you have reached this point of the book, I am hoping you now have a very clear understanding that commercial drones are powerful management tool for your agribusiness.

Part 7:
Yes, But Can I Do This Myself?

Part 7:
Yes, But Can I Do This Myself?

'Farmers need to be able to look closely at the pros and cons for each approach before making a decision, but growers also need to ensure they know what they want to do with a drone before they even consider such details.' [28]

Before I explain the next steps you can take to start introducing the benefits of agricultural drone surveying to your farming business, I want to address the questions we often hear:

- Can't I do this myself?
- Can't I do this more cheaply?
- Why should I appoint others to fly for me when I could purchase my own drone?

These are not unreasonable questions.

Whether it makes sense for you to purchase your own drone(s) or utilise a professional, qualified and insured service provider, without taking on the costs and risks associated with having your own set up, is a major decision that is personal to your circumstances.

However, there are some very important consequences you must consider if you want to reap the full scope of benefits I have described thus far.

By the way, I am not here to convince you to use our services. But I do want you to be clear on what it may mean if you choose to fly for yourself.

OF COURSE YOU CAN, BUT...

There are already many farmers around the world flying their own drones. Drone manufacturers and online software platforms are targeting farmers directly with affordable easy to use agriculture specific bundles comprising drones, cameras and software specific to agricultural applications.

So our answer would be *yes,* you *can* survey your own farm.

But our immediate question back to you is:

- Are you really going to solve your key problems described in this book if you purchase and fly a drone for yourself?
- Are you really optimizing the powerful benefits of smart farming and their impacts on your yields and financial returns I've explained if you do this for yourself?

Do You Do Your Own Dentistry?

As a very silly example, all the tools you need to perform a dental procedure such as replacing a filling or root canal work are freely available for purchase. And I am sure there are hundreds of *YouTube* tutorials that can teach you dentistry. But your dentist is qualified, licensed, insured and importantly, experienced.

You may even find a mate who is handy with a power drill and super glue!

But would you truly entrust such an operation to an unqualified, unlicensed, uninsured and ill-experienced person?

Of course, the answer is no. The same principle applies with agricultural drone surveying.

Importantly, in many countries, there is only so much you can achieve as a *private or recreational* drone user. In South Africa, as well as other jurisdictions there are limits on: *how* drones can be used; *where* drones can be used; *when* they can be used; the distances they can fly from the operator; and, *who* can operate a drone.

Licensing extends these limitations on the principle you have been properly trained, and operate with approved procedures and are insured.

In addition, you will be regularly audited to prove you have upheld the requirements accordingly.

Infringement of the South African regulations *as a licensed operator* carries the risk of any combination of being grounded until the authorities can be assured you have addressed the infringements properly; hefty fines and/or court action.

Legal consequences equally await those *recreational*

users who exceed their legal limitations, especially where you use a drone for any commercial outcome with or without charging for their services, time or data.

In other words, using a drone for your business will still represent a commercial use. This definition of commercial use also applies in Europe.

DIY Drone Surveys

So, let's assume you decide to opt for undertaking your own drone surveys and can legally do so – why should this be counter-productive?

In Part 1, I explained the three core problem areas facing farmers, including: not having accurate information at the right time for effective decision making; having insufficient time to be truly productive; and having inefficient management strategies in place.

Even if I give you a full list of our equipment and software, I am pretty sure you will not gain anything remotely close to the complete scope of outcomes we achieve. You will need to be prepared to invest an enormous amount of effort and money into purchasing the equipment and then learning how to use the equipment, software, interpret and convert the data to do this.

Worse, if you purchase and fly your own drone, then there is a very real risk that you may not be solving your core problems as effectively as you might initially think. I strongly urge you to carefully consider the following points before you go down the route of purchasing and using your own drone.

These are not random arguments; some are based on our experiences setting up our own professional licensed operations.

What are you trying to accomplish and what information will you need to do so?

You must first ask yourself what you truly want to accomplish with the information you collect, because you may not actually be able to obtain all the soil or crop data you need, unless you're prepared to invest time and money purchasing the right drone, equipment and online processing software access.

Drones, the specialist cameras, the flight planning and analytical software are all high-tech products of the digital age and, just like your own laptop, tablet or smart phone, they are constantly updating and improving.

Assuming you make the purchase and spend the time acquiring the skills to fly and use the software, are you prepared to spend time and money keeping pace with the rapidly evolving technology and the software packages?

The capabilities of drones, the sensors and software have already changed significantly in the last three years alone. Given the increasing trend in global investments in drone solutions, these capabilities are only going to escalate. This is all very exciting, but can you afford the time to keep pace with this level and rate of change to acquire the full scope of real-time data you need?

This question goes back to the three core problems and the three solutions provided by commercial drones, the specialist cameras and software discussed in the previous sections of this book.

Does it actually make sense for you to have your own setup?

In other words, if you do this for yourself: will you know all there is to know about your farm, vineyard or orchard; will you really save the time you need to save; and will you really obtain the full range of management outputs from the data collected to increase your yields and profitability?

Can you really afford to spend the time doing this yourself?

Your next question is ,'*What do I want to accomplish with the drone and the information I want to collect?*' because you may not have the time to collect, process, interpret and use your own information.

For starters, you need to spend time researching and purchasing the right drone.

As mentioned in Part 3, there are different drones and cameras for different purposes. Up to a certain degree the footage collected by the *DJI Phantom* series with visible light cameras can have multiple uses but there are limitations compared to using a specialist commercial drone and alternative spectral cameras.

Additionally, you might benefit more from using a fixed wing drone such as the *Sensefly eBee SQ* rather than a multi-rotor drone. You obviously need to make the right choice to suit your needs and budget (see below) to avoid making costly mistakes - such as purchasing a more advanced drone than is actually necessary or not up to the job.

After you have purchased your drone, you must spend time learning to safely operate and fly the drone (and preferably, without crashing).

The rapid rise in drone use has been largely facilitated by the technology that makes drones easier to control and fly. Most drones have '*automation and redundant system capabilities [which] have been built into these [drone] systems to make flying a UAV [drone] much simpler.*' [30]

This means many drones are sold on the principle they can be flown *straight out of the box*.

The popular *DJI Phantom* series includes an automatic take-off and landing feature as well as a *Return to Home* control.

Most of the separate flight planning software also enables autonomous flying and data capture. On top of this, each new generation of the drone invariably includes an upgrade on further improving automatic flying skills. For example, the new *DJI Phantom 4s* incorporate anti-collision sensors and flight software which were not featured in the *Phantom 3* series.

These features can give you a false sense of confidence in your flying skills!

Most professional operators will tell you that it's not a case of *if you crash but when you crash! Googling drone crashes* will give you countless examples of drones crashing on their maiden flights.

Crashing is one thing. What you crash *into* is another consideration. Professional licensed operators typically are required to have emergency procedures in place and carry first aid kits and fire extinguishers - at a minimum.

It is worth bearing in mind the manuals do not tell you all the potential problems you may encounter or how to solve them – for example, what to do when you screen goes blank while you are flying.   For this, you must either go online to search and view the many *How to* videos available or learn the hard way.

As with any software, you need time to become familiar with the programme and learn which composites, indices and automatic reporting functions you should use and when. Most software these days requires you to review online video tutorials and/or engage on dedicated chat groups and forums to resolve your problems, which is time consuming in itself.

The software platform service providers have excellent technical support including answers to *Frequently Asked Questions (FAQ)*, tutorials or a technical person to respond to your specific issues.

However, any helpful guidance they offer to your specific problem can be delayed by lengthy time differences depending on your location.

Some of the top-flight planning and analytical software platforms are situated in the USA, including *Drone Deploy, Precision Hawk* and *Atlas*.

Similarly, the multi-spectral and thermal imaging cameras I have discussed are also manufactured in the USA. This is fine if you live in the USA but you can expect a minimum nine-hour difference if you're in South Africa contacting a technical support team based on the west coast of the USA. In this instance, the earliest you may receive a response to your query would be nine hours after submission.

Managing the transfer of data from the memory cards to your computer for uploading onto the online software programmes takes time. The uploading process itself is time consuming depends on your access to the internet, data transfer speeds and the amount of data you've generated.

Additionally, any editing process of *normal* video footage is usually more time consuming than initially expected. And then there is the time spent downloading, analysing and storing the results for future reference and/or comparison.

So, are you really helping yourself save time and become more productive if you do this for yourself?

If anything, you're *adding* more demands on your time.

Best case, you will be learning to improve your flying skills and software competency through a process of trial and error without damaging or losing your drone. Worst case you will severely damage and/or lose your aircraft with this hit or miss approach; with all the knock-on ramifications that may follow.

Are you prepared to invest the money?

Again, your specific requirements will determine how much you will need to spend purchasing the correct drone, cameras and access to the online flight planning and processing software. There are many hidden costs to purchasing and operating the whole system for yourself.

I recently saw a visual representation of the cost of establishing and owning a commercial drone operation. [31]

It drew parallels with the dimensions of an iceberg.

It is a well-known fact that for any iceberg, you will typically find over 90% of the iceberg lies *below* the waterline.

It was suggested the same principle applies to commercial drone operations insofar as hidden costs amount to 85% of the total investment made. In other words, for every $1 planned expense, there is another $6 unplanned hidden but essential cost. While I acknowledge you may not be planning to establish a full commercial operation I know from our own experience you will incur hidden costs that cannot be avoided including the cost of your time.

It is also important to bear in mind you will need different equipment for different times during a growing season and, at the same time, you may also find this expensive equipment will be lying idle for much of a season.

This is a particularly wasteful investment if you consider when, for example, the camera becomes obsolete or new cameras offer better economic advantages for your business.

You are also paying for subscriptions to online processing platforms for the occasional data set. There are month-on-month payment options, but these are more expensive than an annual subscription and it is likely you will not obtain a full value for money from the subscription. It makes sense for us because we are the full-time service provider.

Perhaps you are thinking that you could *resell* the equipment later on if you do not use it much after all.

I'm afraid, not so.

Much like purchasing a new vehicle, you can expect an immediate depreciation of your drone setup once you've made the purchase. This is exacerbated by the fast pace at which manufacturers are turning out more advanced products and by the subsequent drop in price of the *older* models. Your drone setup will typically lose approximately 4% of its value per month over two years and you may find you can only give away your expensive purchase. [32]

The costs associated with a drone purchase are entirely dependent on what you want to achieve.

It you just want to observe your operations or take simple photographs and video footage you could use anything from a *DJI Mavic Air* to a *Phantom 4 Pro*. Prices currently range from $900 through to $1,500 depending on where you purchase the drone.

Alternatively, the popular fixed wing option, the *Sensefly eBee SQ,* starts from $12,000. If you want to start adding multispectral cameras, the new *MicaSense RedEdge M* camera carries a price tag in the order of $4,900.

The *Flir Zenmuse XT* thermal imaging camera for the *DJI Inspire 1* can set you back in the order of $6,000 to $11,750 depending on the final camera specifications chosen.

If you're contemplating using drones to monitor your crop health then you may wish to purchase a drone that either has a fitted specialist camera as standard, enables the switching of cameras/sensors or can have a second camera attached simultaneously.

Either of these options will have additional cost implications that it will be best to consider from the start.

Retro-fitting specialist cameras to some drones is possible but this is costly, can reduce the performance of the drone or is just not possible at all. A mount for the *MicaSense RedEdge M* for the *DJI Phantom 4 Pro* may set you back $385 depending on the supplier.

Footage generated by standard visible spectrum cameras can be used to create plant health maps, however these maps are not as comprehensively *revealing* as those generated by the more expensive multispectral cameras.

Another alternative to purchasing a specialist camera is to modify a standard camera to capture near infrared using filters. However, 'image quality is inconsistent across manufacturers meaning some may lead to poor map quality' and these modified cameras are still expensive at between $1,200 - $7,000. [33]

If you require good quality crop information, then specialist cameras are ultimately the better option as explained in Part 3.

There are dedicated drone packages that are pre-fitted with or designed to carry specialist cameras for agricultural use. One such commercial package offered by *Precision Hawk* and featuring the *DJI Phantom 4 Pro* will cost in the order of $2,000. This price includes one year's access to their analytical software, however this package does not include a specialist camera. There are options to upgrade the *DJI Phantom 4 Pro* to carry the *Sentera NDVI* camera for an additional approximately $2,000. Other specialist packages can set you back from $8,000 upwards for multirotors and starting from $12,000 for fixed wing options.

You will need to purchase at least one subscription to an online software platform for accurate flight planning for the surveys, processing and rendering maps from the data collected by the drone. These subscriptions range in cost from $49/month (Atlas), $149/month (Pix4DAG) to $83/month (DroneDeploy) and will provide access to certain tools and features only. More expensive *unlimited* options are available at $99/month (Atlas) and $249/month (DroneDeploy) where discounts for annual subscriptions apply.

Alternatively, Drone Deploy offers a comprehensive *Precision Agriculture Package* at $999 per annum. This may not affect you but some of the software plans will limit the number of users that can access an account.

You can sign up for free access on some of these analytical software programmes but your activities are usually limited insofar as: images may be of a lower quality; the turnaround time for processing is not as quick as a paid subscription; you may not be able to generate all the maps you require or access all the apps and conversion tools; and there are restrictions on the number of images that can be uploaded per month. Also, some free subscriptions are only available for a limited time such as just seven days.

Then you will almost certainly need to purchase additional peripherals such as extra batteries if you want to fly over large areas at one time. Batteries are expensive. They are normally in the order of $170 – $199 for the *Phantom 4 Pro* and *Inspire 1* respectively, especially if you purchase extra batteries after the initial purchase.

Unless you commit to a more expensive/advanced commercial drone from the start (which I would not recommend from experience) you are generally *stuck* with the specifications and capabilities of your original purchase which suited your needs at that time.

I guarantee that once you start using a drone, you will soon want to expand its application across your farm, and almost certainly start moving beyond the limits of your current equipment (and software capabilities).

If this is the case, you will need to upgrade.

You will need to be prepared to keep pace with the rapidly evolving developments in commercial drones, cameras and processing software and *add-ons* if you want to keep maximising the leverage gained for your farming operations with your setup.

This will also carry an expense.

It is also wise to insure yourself to cover yourself against loss or damage to the drone on your property and/or third-party insurance if you really want to protect yourself where there are neighbouring properties at risk from an errant drone!

As a point of note, you may encounter a developing trend in many countries where insurance brokers will *not* cover you unless you have a formal remote pilots license, at a minimum.

This requirement adds a whole different level of time and budget commitments which should not be underestimated.

The fee for obtaining a *Remote Pilot License (RPL)* in South Africa (from one of only four *SACAA* approved schools) is approximately $2,300 (excluding the cost of your time, any travel and accommodation). You must pass a theoretical and practical examination on a range of topics as well as pass a *Class 4* medical examination and obtain a *Restricted Radio License*.

Incidentally, if your use of a drone on your farm constitutes a *commercial use* (as per South Africa and Europe) then obtaining an RPL will not be enough. You will need to become a licensed and insured operation as well.

At the end of the day, there is a real risk the apparent savings you think you are making by purchasing your own drone setup are a false economy. This false economy will become worse if the legal requirements for drone use mean your own uses on your farm constitute a commercial operation and must therefore be licensed – but this is the topic for a whole new book and I will not delve into the subject here. Suffice to say, the commercial licensing process is not a process to be taken lightly or ignored. This is especially the case in South Africa.

Can you really afford to get it wrong?

I think I have already covered this question in considerable detail. However, here is a summary of the key costs I have included in this chapter. [34]

DJI Phantom 4 Pro	$1,500
Sensefly eBee SQ	$12,000 +
Micasense Rededge M camera	$4,900
Flir Zenmuse XT thermal imaging camera	$6,000 - $11,750
Mount for the Micasense Rededge M camera for the DJI Phantom 4 Pro	$385
Commercial Agricultural Drone and Software Package	$2,000
Commercial Agricultural Drone and Software Package & Sentera NDVI camera	$4,000
Software Subscription – Atlas	$49/month
Software Subscription – Pix4 DAG	$149/month
Software Subscription – DroneDeploy	$83/month
Precision Agriculture Software Package – Drone Deploy	R999/year
Extra Batteries – DJI Phantom 4	$170 each
Remote Pilot License Course (South Africa)	$2,300 +
Insurances	$ Quote Required

This list includes brands and models we currently utilise (July 2018). Hardware is constantly evolving which means we will update models or switch brands as better options become available.

By now, you should appreciate commercial drone surveys are part and parcel of precision farming and that, in turn, precision or smart farming is transforming agriculture. If you truly want dependable, consistent, accurate, repeatable results and analysis for your farm, vineyard, orchard or plantation, then you will need qualified and licensed professional services.

SHOULD I USE PROFESSIONALS?

Yes. Why divert what time and resources you have from your farming business when there are fully insured, trained and licensed professional operators that have acquired the right drones, equipment, software and high standard of skills to assist you?

If you have got this far in the book, you will now know that using commercial drone surveying in agriculture is more than just flying and taking photographs over your fields, vineyards or orchards.

There is far more than meets the eye with this exciting, versatile and rapidly evolving technology. If you want to achieve the most impact by streamlining and optimising your farming operations for higher yields and profits, then let the professionals assist you.

The Legal Argument

As I've mentioned, a very important reason for using professionals is that using your own drone on your farm may breach your country's civil aviation regulations pertaining to private or recreational drone use. Again, using your own drone to support your farm business constitutes a commercial use in South Africa alone and may do so elsewhere.

As it stands, different national jurisdictions have different civil aviation regulations and South Africa has some of the toughest regulations concerning drone use. But one thing all jurisdictions share in common is that breaching safety and air space regulations is treated very seriously.

For example, a violation of the South African regulations may occur where: you're operating too close to public roads; flying farther than 500m from yourself; flying higher than 120m above ground level; using a drone to spray crops; or flying within controlled airspace for manned aviation.

Bear in mind, pleading ignorance will not help you. You will run a serious risk of prosecution and fines, at a minimum. First time fines start at $3,100!

Prison sentences of up to ten years are also a very real possibility. [35]

At the time of finishing this book the UK Government were introducing on-the-spot-fines of $3,300 for breaking new restrictions (July 2018).

And should you have an accident with your drone that results in injury or death, property damage, fire or collision with manned aircraft then the legal consequences are far worse. You will be held criminally negligent as well as subject to fines and/or prison sentences.

Can you really afford to take this risk?

'But I Have a Friend with a Drone…'

Perhaps you know someone who has purchased and turned their hand to flying a low-cost recreational drone with a camera and they've offered to fly over your farm for mate's rates. Where's the harm in this?

Again, I would caution you about following this route because of the legal repercussions alone.

First, your friend is almost certainly operating illegally in many countries, including South Africa. Receiving remuneration for their time or the footage they collect immediately renders them a commercial operator and therefore subject to licensing. You will also be exposed to legal repercussions whether you pay for the footage or not. If the illegal operator you have commissioned has an accident while flying on your property, *you* will still be held criminally negligent.

Are you really prepared to run this risk?

Second, you are unlikely to reap the full scope of benefits described in this book from amateur operators. Amateur operators will not have the training, qualifications, the staff, the stringent safeguards, the appropriate CAA licenses and professional indemnity insurances in place.

These requirements are there to: promote safety; to hold us accountable for the quality of the professional services we provide; and ultimately protect you.

If you choose to use an amateur then you risk wasting time and money on what will constitute an inferior service despite their best intentions. It will be an inferior service primarily because they are unlikely to have the full suite of equipment or skills that constitute a professional service.

And then there is the potential for costly errors in judgment, delays or unintended outcomes occurring with limited or no recourse you can pursue thereafter.

Rather than risk avoidable legal contraventions, placing others' property or safety in the air and on the ground at risk, or experiencing a poor-quality service, find and let the right people help you.

It may appear I am throwing DIY under the bus for my own gain. This is not my intention. Professional drone operators take safety and civil aviation obligations seriously. Importantly, so should you.

Part 8: Making It Happen

Part 8: Making It Happen

'Agricultural drones are here to stay. Farmers who embrace the technology and integrate it into their precision programs will wonder how they ever got along without it.' [36]

I want to see this innovative technology spread and for all farmers to experience the benefits from introducing precision farming in to their operations and their financial situations.

Drone surveying *'allows farmers to move from the intuitive decision making to analytical decision making.'* [37]

Don't get left behind in the smart farming transformation.

We can help integrate the value of agricultural drone services into your operations smoothly and efficiently. We will help you farm smarter, achieve higher yields more sustainably and acquire those *painless profits* that currently seem out of reach.

If you want to use your own drones yourself on your farm, then there are a number of steps you can take.

Civil Aviation Requirements

First and foremost, check if you are able to use drones on your farm as part of your business or in the *airspace* above your farm without licensing. Consult your national civil aviation authority representatives or website or contact other professionals such as aviation lawyers or your local manned aviation flying school.

There are other websites which summarise global drone laws such as *http://droneregulations.info*

I recommend you still check within your country in case there have been updates that are not reflected on these websites.

Desk Search

A *desk search* simply means spending time on the computer searching. I recommend that you start *Googling* and *YouTubing* on a regular basis. Read as many articles as you can about commercial drones and agribusiness.

Start to improve your knowledge of the topic. Especially read anything that might be related to your farm and your crops .. and drone surveys. Combine those elements in your search terms.

Try and pick the latest reports and articles. Search for blogs and podcasts and videos

Search Associations

If you belong to an agribusiness cooperative or a professional association check if they are offering information about drones.

Search Professional Bodies

More often than not we discover that a local organisation or professional body is offering workshops and training in agribusiness that feature a drone component.

You might be surprised that your local *Chamber of Commerce* is offering an agribusiness workshop that features drone technology.

Search Government Bodies

Search for government initiatives that relate to precision farming or agribusiness development. Most governments offer an amazing range of courses, contacts and even subsidised training. I recently saw grant funding for GPS farming technology! Sometimes research grants are available from universities looking to pilot new technology. They are looking for guinea pigs! Most farmers say they are too busy to search but then get annoyed when they discover their neighbour accessed a round of funding!

Search Funding Opportunities

As I mentioned, often governments, companies and universities - banks even - have funding for trials and often those funds go begging.

The advice here is to dedicate time to desk searches.

Read Case Studies

We've included several brief case studies to show what's possible. Again, read broadly. Include a blend of local and international case studies. Simply broadening your knowledge of drones and agribusiness will help clarify your own objectives.

Seek Professional Help

I hope I have already made a strong case on why you should consider using professional operators rather than friends, amateurs or illegal operators on your farm.

But just to state one last time, please don't be tempted by the cheaper or easier offer you may get or even by deliberately operating illegally yourself. There are serious implications that may ultimately affect your farm and livelihood.

Instead, do your research and make sure you can operate legally (if this suits your needs) or find the right licensed and insured professionals to help you instead.

Talk To Us

Of course, please call us.

For Readers *inside* South Africa

Please note at the time of printing we are finalising our drone operating licensing and permits with the *South African Civil Aviation Authority (SACAA)*.

Louise is already a licensed drone *pilot* with over 300 flying hours. She was one of the first women to obtain her pilot license from the *South African Civil Aviation Authority* in 2016.

Terreco Aviation's intention is to be the premier provider of *commercial drone* surveys in South Africa. We will commence providing a full range of drone services once we have completed the licensing process in 2019.

For reasons outlined in the book you should only deal with licensed pilots and service providers.

For Readers *outside* South Africa

We are able to quote and provide our services outside South Africa. If you're ready for smarter farming your next steps are:

- Read case studies from other parts of the world on www.terrecoaviation.co.za/casestudies. We will add our own case studies once we are fully licensed (due 2019).
- Watch for our webinars on our website and YouTube channel – www.YouTube.com/TerrecoAviation. We will be running monthly webinars where we will continue to talk about the use of drones in agriculture.
- Request a presentation at your conference, company or association meeting.
- Come and have a chat with us – in person or via video chat on Skye, Zoom or FaceTime.

Other contact details for Terreco Aviation:

Email: info@terrecoaviation.co.za

Webpage: www.terrecoaviation.co.za

Twitter: @TerrecoAviation

Facebook: http://fb.com/TerrecoAviation

Instagram: @TerrecoAviation

YouTube: TerrecoAviation

Case Studies

The following comprises a selection of summarised published case studies and articles highlighting the benefits of commercial drone services as noted by farmers and growers themselves.

CROP AND TIME SAVING WEED MANAGEMENT [38]

The easy arrangement and deployment of a professional drone service provider helped a coffee grower in the USA respond to a fast growing weed infestation more quickly than using traditional manned aircraft.

The grower noted that the normal turnaround time for hiring an aircraft to review and assess the crop damage from the weed was two weeks and by this time the problem would have significantly escalated.

The professional drone services provider was able to respond and provide the required information within 48 hours. This represented both a significant saving for the grower in that fewer plants were lost and cost savings were made with the treatment of the infected areas, alone.

PLANT COUNTING [39]

A tomato grower in the USA wanted to confirm the number of plants that had been transplanted by another company onto a 74-acre field (30ha). As the grower is charged per transplanted plant, it was important to have an accurate count. Traditional methods involve making a rough count over a selected area and creating an average for the entire field. An estimate from greenhouse plants, indicated the tomato plant loss would be in the region of approximately 5%. A 30-minute professional drone survey and plant count analysis revealed an actual 26% loss of the plants the farmer had purchased.

Armed with this information, the farmer was in a far better position to go back to the planter and/or insurers to recover the losses.

PLANT DAMAGE ASSESSMENT [40]

Another case study shows the value of drone sourced information for insurance claims after heavy rains. The data collected and mapped for a 100-acre field (40 hectares) by a professional service helped a farmer recoup an additional $110,000 for crop losses after the drone mapping convinced the insurance inspector to re-assess a field.

WEED INVASION ASSESSMENT [41]

Professional drones were instrumental in helping a farmer make big savings in labour to contain and remove an aggressive weed before the infestation became a runaway problem. Drone surveys identify small patches of the weed (which are otherwise difficult to see) using plant health maps.

REPLANT OR NOT [42]

Commercial drone services helped to inform a farmer on whether he should replant his fields.

A farmer had completed his own field assessment on foot to review stand loss but was unsure if it still made economic sense to replant. A 17-minute drone survey of his 80-acre site (32 ha) produced a stand count, stand loss percentage and weed pressure percentage report within 24 hours. On the basis of the report and associated maps, the farmer was able to confirm the potential loss in harvest revenue would not justify the time and cost of replanting.

INCREASE IN CROP YIELDS [43]

A case study in France illustrates the value of drones in terms of boosting crop yields through effective crop intelligence gathering. Commercial drones have been mapping over 7,000 ha (17,297 acres) of oil seed rape and cereals for a French farming cooperative at selected intervals since 2015.

Processed images and maps are used to guide fertilization programmes and determine variable rate applications.

The farmers who have used these services have recorded an increase in their average yields by 10% compared to those areas where drones are not used.

INCREASING PRODUCTIVITY [44]

Contour mapping using drone services proved to be 75% more efficient than traditional methods for creating a planting plan for sugar cane producers in Brazil.

Data was collected in four hours compared to the standard two days and data was used to elevation maps, 3D representations of the fields and ultimately a planting plan.

EARLIER DETECTION WITH MULTISPECTRAL CAMERAS [45]

Case studies show how multispectral cameras that can detect the red edge spectral band provide an additional advantage for plant stress detection.

The red edge spectral band gives an extra edge in detecting plant stress because it can be used with additional indices and mapping options, such as the chlorophyll map. This means plant problems can be detected earlier compared to standard cameras or non-red edge detecting multispectral cameras where the consequences of the stress are still invisible to NDVI.

'For many types of crop stress such as water, nitrogen, and fungal diseases, early detection and fast response is key to prevent yield loss and maintain maximum potential. A sensor without the red edge band may still enable the detection of stress and disease, but it may be after there is a chance to recover yield potential which would result in a reduction of profit per acre.'

References

Introduction

1. PwC (2016). 'Clarity from Above. PwC Global Report on the Commercial Applications of Drone Technology.' (www.dronepoweredsolutions.com)

2. Brodie C (2018). 'Why the world's fastest-growing populations are in the Middle East and Africa. World Economic Forum.

3. WWF (undated). 'Agriculture: Facts and Trends South Africa.'

4. PwC (2016). 'Clarity from Above. PwC Global Report on the Commercial Applications of Drone Technology.' (www.dronepoweredsolutions.com)

5. Letsebe K (2017). 'Global Drone-Powered Solutions valued at $127bn: Report.' (www.itweb.co.za).

6. PwC (2016). 'Clarity from Above. PwC Global Report on the Commercial Applications of Drone Technology.' (www.dronepoweredsolutions.com)

Part 1: Farming Today

7. Doering C (2014). 'Growing Use of Drones Poised to Transform Agriculture.' USA Today.

8. Mazur M (2016). Six Ways Drones are Revolutionising Agriculture.' Technology Review (https://www.technoogyreview.com)

9. PwC (2016). 'Clarity from Above. PwC Global Report on the Commercial Applications of Drone Technology.' (www.dronepoweredsolutions.com)

Part 2: What We See That Works For Farmers

10. European Parliament (2014). 'Precision Agriculture: An opportunity for EU Farmers - Potential Support with the Cap 2014-2020.' Directorate-General for Internal Policies. Policy Department B Structural and Cohesion Policies. Agricultural and Rural Development.

11. Mazur M (2016). 'Six Ways Drones are Revolutionising Agriculture.' Technology Review (www.technologyeview.com/)

Part 3: Having Better Quality Information

12. Karpowicz J (2016). 'Above the Field with UAVs in Precision Agriculture: Discover How and Why UAVs are set to Impact and Change the Way Precision Agriculture Professional Operate.' Commercial UAV Expo.

13. Parrot (2016) Parrot Sequoia: Capture the Invisible.' (www.parrot.com).

14. Electrophysics Infrared Inspection (2011) 'Infrared Inspection White Paper: Understanding Infrared Camera Thermal Image Quality.' Groupe Sofradir.

15. Karpowicz J (2016). 'Above the Field with UAVs in Precision Agriculture: Discover How and Why UAVs are set to Impact and Change the Way Precision Agriculture Professional Operate.' Commercial UAV Expo.

16. MicaSense (2017). 'An Overview of the available layers and indices in Atlas.' (https://support.micasense.com)

17. MicaSense (Undated). 'The Science Behind MicaSense.' (https://blog.micasense.com)

18. Drone Deploy (undated). 'Crop Scouting with Drones Identifying Crop Variability with UAVs: A Guide to Evaluating Plant Health and Detecting Crop Stress with Drone Data.'

19. McKinnon T Dr (2016). 'Agricultural Drones: What Farmers Need to Know.' Agribotix

Part 4: Spend More Time On Productive Tasks

20. Nixon A (2017). 'Best Drones for Agriculture 2017: The Ultimate Buyer's Guide.' (www.thebestdroneforthejob.com)

21. Karpowicz J (2017). 'How are Precision Agriculture Professionals Using Drones in 2017?' Commercial UAV Expo.

22. Drone Deploy (2017). 'From Pre-Planting to Post Harvest: Harnessing the Power of Drones Year Round.' (https://blog.dronedeploy.com)

Part 5: Creating Better Management Strategies

23. Nixon A (2017). 'Best Drones for Agriculture 2017: The Ultimate Buyer's Guide.' (www.thebestdroneforthejob.com)

24. Nixon A (2017). 'Best Drones for Agriculture 2017: The Ultimate Buyer's Guide.' (www.thebestdroneforthejob.com)

25. Drone Deploy (2017). 'From Pre-Planting to Post-Harvest: Harnessing the Power of Drones Year-Round.' (https://blog.dronedeploy.com).

Part 6: The 5-Step Precision Framing Blueprint to Smarter Farming

26. South African regulations do not permit 'beyond visual line of sight' (or BVLOS) flying for licensed operators yet. 'Extended visual line of sight' (EVLOS) is an option for licensed operators provided with suitably trained observers to assist the pilot.

27. Parrot (2016) 'Parrot Sequoia: Capture the Invisible.' (www.parrot.com).

Part 7: Can I Do This Myself?

28. Karpowicz J (2016). 'Above the Field with UAVs in Precision Agriculture: Discover How and Why UAVs are Set to Impact and Change the Way Precision Agriculture Professionals Operate.' Commercial UAV Expo.

29. Karpowicz J (2017). 'Seven Commercial Drone Predictions for 2017.' Commercial UAV Expo.

30. Karpowicz J (2016). ' Above the Field with UAVs in Precision Agriculture.' Commercial UAV Expo.

31. Chapman A (2017). 'Drones: Total Cost of Ownership (TCO).' Australian UAV.

32. Chapman A (2017). 'Drones: Total Cost of Ownership (TCO).' Australian UAV.

33. Drone Deploy (2017). 'Crop Scouting with Drones Identifying Crop Variability with UAVs: A Guide to Evaluating Plant Health and Detecting Crop Stress with Drone Data.'

34. All listed costs were correct at the time of printing (July 2018).

35. Kent and Co (2018) Get Legal. Get Airborne. Simple. (www.kentand.co.za/RPAS/).

Part 8: Making It Happen

36. McKinnon T Dr (2016) 'Agricultural Drones: what farmers need to know.' Agribotix.

37. Quebrajo L, Perez-Ruiz M, Perez-Urrestaraz L, Martinez G and Egea G (2017). Linking Thermal Imaging and Soil Remote Sensing to Enhance Irrigation Management of Sugar Beet.' Biosystems Engineering.

Case Studies

38. Drone Deploy (2018) 'Drones in Agriculture – The Ultimate Guide to Putting Your Drone to Work on the Farm.'

39. Drone Deploy (2018) 'Drones in Agriculture – The Ultimate Guide to Putting Your Drone to Work on the Farm.'

40. Drone Deploy (2017). 'Seven Ways to Use Drone Mapping on the Farm this Season.' (https://blog.dronedeploy.com)

41. Drone Deploy (2017). 'Seven Ways to Use Drone Mapping on the Farm this Season.' (https://blog.dronedeploy.com)

42. Drone Deploy (2018). 'Drones in Agriculture – The Ultimate Guide to Putting Your Drone to Work on the Farm.'

43. SenseFly (2016). 'Flying High – How a French farming cooperative used drones to boost its members' crop yields.'

44. Drone Deploy (2018). 'Drones in Agriculture – The Ultimate Guide to Putting Your Drone to Work on the Farm.'

45. MicaSense (2018) Detecting Disease Earlier: The Importance of the Red Edge Band. (https://blog.micasense.com).

Bibliography

- Brodie C (2018). 'Why the world's fastest-growing populations are in the Middle East and Africa.' World Economic Forum.
- Chapman A (2017). 'Drones: Total Cost of Ownership (TCO).' Australian UAV.
- Doering C (2014). 'Growing Use of Drones Poised to Transform Agriculture.' USA Today.
- Drone Deploy (undated). 'Crop Scouting with Drones. Identifying Crop Variability with UAVs: A Guide to Evaluating Plant Health and Detecting Crop Stress with Drone Data.'
- Drone Deploy (2017). 'From Pre-Planning to Post Harvest: Harnessing the Power of Drones Year-Round.' (https://blog.dronedeploy.com)
- Drone Deploy (2018). 'Drones in Agriculture – The Ultimate Guide to Putting Your Drone to Work on the Farm.'
- Drone Deploy (2017). 'Seven Ways to Use Drone Mapping on the Farm this Season.' (Https://blog.dronedeploy.com).
- Electrophysics Infrared Inspection (2011) 'Infrared Inspection White Paper: Understanding Infrared Camera Thermal Image Quality.' Groupe Sofradir.
- European Parliament (2014). 'Precision Agriculture: An Opportunity for EU Farmers – Potential Support with the CAP 2014-2020.' Directorate-General for Internal Policies. Policy Department B Structural & Cohesion Policies. Agriculture and Rural Development.

- Karpowicz J (2016). 'Above the Field with UAVs in Precision Agriculture: Discover How and Why UAVs are Set to Impact and Change the Way Precision Agriculture Professionals Operate.' Commercial UAV Expo.
- Karpowicz J (2017). 'How are Precision Agriculture Professionals Using Drones in 2017?' Commercial UAV Expo.
- Karpowicz J (2017). 'Seven Commercial Drone Predictions for 2017'. Commercial UAV Expo.
- Kent and Co (2018). Get Legal. Get Airborne. Simple (www.kentand.co.za/RPAS/)
- Letsebe K (2017) 'Global Drone-Powered Solutions valued at $127 Bn: Report' (www.itweb.co.za)
- Mazur M (2016). Six Ways Drones are Revolutionising Agriculture.' Technology Review (https://www.technoogyreview.com).
- McKinnon T, Dr (2016). 'Agricultural Drones: What Farmers Need to Know.' Agribotix.
- MicaSense (2017) 'An Overview of the Available Layers and Indices in Atlas.' (https://support.micasense.com).
- MicaSense (undated) 'The Science Behind MicaSense.' (https://blog.micasense.com)
- MicaSense (2018). 'Detecting Disease Earlier: The Importance of the Red Edge Band.' (https://blog.micasense.com)
- Nixon A (2017) 'Best Drones for Agriculture 2017: The Ultimate Buyer's Guide.' (www.thebestdroneforthejob.com)
- Parrot (2016) Parrot Sequoia: Capture the Invisible.' (www.parrot.com).
- Precision Hawk Case Study (undated): 'Timely and Accurate Damage Assessment for Drowned Tobacco.' (www.precisionhawk.com).
- PwC (2016). 'Clarity from Above. PwC Global Report on the Commercial Applications of Drone Technology.' (www.dronepoweredsolutions.com).

- Quebrajo L, Perez-Ruiz M, Perez-Urrestaraz L, Martinez G and Egea G (2017). 'Linking Thermal Imaging and Soil Remote Sensing to Enhance Irrigation Management of Sugar Beet.' Biosystems Engineering.
- Sensefly (2016). 'Flying High – How a French Farming Cooperative Used Drones to Boost its Members' Crop Yields.'
- WWF (undated). 'Agriculture: Facts and Trends South Africa.'

Glossary

The following acronyms and abbreviations have been used in this book.

2D	Two dimensional
3D	Three dimensional
BVLOS	Beyond visual line of sight
CIR	Colour infrared composite
DSM	Digital surface model
EVLOS	Extended Visual Line of Sight
FAQ	Frequently asked questions
GIS	Geographical information system
GPS	Global positioning system
K	potassium
.kmz	Keyhole Mark-up language Zipped - file extension for a placemark file used by *Google Earth*
LiDAR	Light detection and ranging
N	nitrogen
NDRE	Normalized difference red edge
NDVI	Normalized difference vegetation index
OSAVI	Optimized self-adjusted vegetation index
.pdf	Portable Document File – a file format that presents a printed document as an electronic image
RGB	Red green blue light wavelengths
RPA	Remotely piloted aircraft

RPAS	Remotely piloted aircraft systems
RPL	Remote pilot license
RPV	Remotely piloted vehicle
S	sulphur
SACAA	South African Civil Aviation Authority
UAVs	Unmanned aerial vehicles
UAS	Unmanned aerial systems
VARI	Visible atmospherically resistant index
VLOS	Visual line of sight
WWF	World Wide Fund for Nature

Acknowledgments

Precision Farming from Above has been ten months in the making but it has been a thoroughly enjoyable journey thanks to so many people who have helped this book become a reality in one way or another.

I would like to sincerely thank:

My wonderful parents for always encouraging and supporting me in all my endeavors wherever they have taken me.

The farmer friends and their families who freely shared their experiences with me and answered my many questions.

The friends and colleagues in the aviation community and drone industry who have helped me acquire the flying skills, knowledge and passion for using drones.

My special group of friends for their unfailing support and encouragement over the last ten months and for forgiving me for regularly turning down invitations or not reciprocating their hospitality while I prepared the book.

The Peer Reviewers who gave their time freely and helped to improve the book's content with their expertise and insightful advice including Tamme van der Wal, Spurgeon Flemington, Duncan Scott and Mike Kearney.

All those who wrote so many kind words in their testimonials that are included at the start of the book, including Duncan Scott, Sharon Rossmark, Professor Filippo Tomasello, Purdy Patel, Rob Somers, Dr GV Price, Dr Mandy Uys, Bruce Fennessy and Barry Dean Delport.

Andrew Priestley, business coach, cartoonist and publisher who has given his time and valuable advice generously and without hesitation from start to finish.

Dent Global for their KPI programme which set me on the road of becoming an author.

Finally I would like to thank Duncan Scott, my steadfast business partner of eight years and counting, and the awesome team at *Terreco* for their support throughout this undertaking.

Note: Every reasonable effort has been made to acknowledge the ownership of the material included in this book.

Any errors in referencing authors that may have occurred are inadvertent and will be corrected in subsequent editions provided notifications are sent to the author. Any errors in the text are my responsibility alone.

About the Author

Louise Jupp has a Master's Degree in Environmental Science and over 26 years experience in environmental management in the UK, Europe and Africa.

Louise is originally from the UK and has been living in South Africa since 2000. She co-founded *Terreco Aviation (Pty) Ltd* with her business partner in 2016. *Terreco Aviation* was established to  provide professional drone-supported consulting services to selected sectors including agriculture.

"We recognised the enormous value complex drone surveying systems provide in terms of generating high quality data at the right time enabling better time management and more effective decision-making."

"We relish the idea we can drive a revolution in traditional agricultural practices using this exciting technology which includes innate positive environmental benefits."

Louise believes drone surveying will be instrumental

in helping farmers improve their yields and achieve their goals in a more sustainable way. Her vision is to see global food production increase with less environmental cost.

Contact the Author

If you have any feedback on this book please feel free to contact me. I would love to hear how commercial drone surveys have turned around your agribusinesses.

You can contact Louise via:

Email: Louise@terrecoaviation.co.za
Website: www.terrecoaviation.co.za
LinkedIn: Louise Jupp LinkedIn

PRECISION FARMING FROM ABOVE

Lightning Source UK Ltd.
Milton Keynes UK
UKHW020611070519
342237UK00014B/1369/P